Taunton's **COMPLETE ILLUSTRATED** *Guide to Using*

Woodworking Tools

LONNIE BIRD

The Taunton Press

The Taunton Press
Inspiration for hands-on living®

The Taunton Press, Inc., 63 South Main Street, PO Box 5506, Newtown, CT 06470-5506
e-mail: tp@taunton.com

Distributed by Publishers Group West

EDITOR: Tony O'Malley
DESIGN: Lori Wendin
LAYOUT: Cathy Cassidy
ILLUSTRATOR: Mario Ferro
PHOTOGRAPHER: Lonnie Bird

LIBRARY OF CONGRESS CATALOGING-IN-PUBLICATION DATA:

Bird, Lonnie.

Taunton's complete illustrated guide to using woodworking tools / Lonnie Bird.

p. cm.

ISBN 1-56158-597-1

1. Woodworking tools. I. Title: Complete illustrated guide to using woodworking tools. II. Taunton Press. III. Title.

TT186.B51723 2004

684'.08--dc22

2004006003

Printed in the United States of America
10 9 8 7 6 5 4 3 2 1

The following manufacturers/names appearing in *Taunton's Complete Illustrated Guide to Using Woodworking Tools* are trademarks:
Lie-Nielsen Toolworks™, Lufkin®, Ohio Tools®, Sargent®, Stanley®, Starrett®, X-Acto®

About Your Safety: Working with wood is inherently dangerous. Using hand or power tools improperly or ignoring safety practices can lead to permanent injury or even death. Don't try to perform operations you learn about here (or elsewhere) unless you're certain they are safe for you. If something about an operation doesn't feel right, don't do it. Look for another way. We want you to enjoy the craft, so please keep safety foremost in your mind whenever you're in the shop.

To my daughters, Rebecca and Sarah, who fill my life with great joy.

Acknowledgments

WRITING A BOOK ISN'T A SOLO ENDEAVOR; there are many people who contribute ideas, vision, and work to make such a project successful.

With that in mind, I wish to express my sincere appreciation to Helen Albert for giving me the opportunity to write this book. Many thanks also to Tony O'Malley, editor, for his hard work and dedicaton on this book project.

Thanks also to those who provided tools for photographs, including Chuck Hardin and Angie Shelton of Delta/Porter-Cable, Scott Box of Powermatic, Todd Walter at Dewalt, Cliff Paddock at CMT, and Tom Lie-Nielsen at Lie-Nielsen Toolworks.

Special thanks to Linda, my wife and life partner, for her patience and help through this entire project.

Contents

PART TWO Benches, Clamps, and Assembly · 44

SECTION 3 | Benches and Clamps · 46

SECTION 4 | Gluing and Assembly · 63

PART THREE Hand Tools · 78

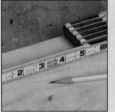    
  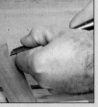

Introduction

THERE ARE FEW ACTIVITIES that provide the enjoyment and deep sense of personal satisfaction as woodworking. As you join, shape, and smooth the wood, you feel growing anticipation as the piece nears completion. The excitement builds as the piece is assembled and the finish applied. The pleasure of woodworking is in the process of using tools to create furniture that will last for several generations.

If you are new to woodworking you may be wondering where to begin. It's always a good idea to start with a few hand tools, such as a couple of planes, a set of chisels, some layout tools, and a handsaw. Using hand tools requires patience and a measure of skill, but in the process you'll learn all about grain direction, accurate layout, and the importance of sharp tools. And, as you learn to cut and fit a dovetail joint or carefully shape the sensuous curves of a table leg, the hand tools will create textures and surfaces that distinctly say, "handmade."

Learning to use power tools can be equally satisfying; woodworking machines provide accuracy and efficiency that's difficult to match with hand tools. The tablesaw is the first power tool that many woodworkers purchase. It can accurately rip and crosscut as well as cut many joints. The jointer and planer are a team that can efficiently flatten and plane lumber to size. And almost every woodworking shop has a bandsaw; it's the tool of choice for cutting curves and the only tool that can resaw bookmatched panels and veneer.

Essentially, power and hand tools are of equal importance: Machines provide efficiency for labor-intensive tasks, such as sawing and planing; hand tools are used to create fine details that machines can't duplicate.

As you peruse the pages of this book, I hope that you'll learn many new skills while experiencing the intense enjoyment that woodworking provides.

How to Use This Book

IRST OF ALL, this book is meant to be used, not put on a shelf to gather dust. It's meant to be pulled out and opened on your bench when you need to do a new or unfamiliar technique. So the first way to use this book is to make sure it's near where you do woodworking.

In the pages that follow you'll find a wide variety of methods that cover the important processes of this area of woodworking. Just as in many other practical areas, in woodworking there are often many ways to get to the same result. Why you choose one method over another depends on several factors:

Time. Are you in a hurry or do you have the leisure to enjoy the quiet that comes with hand tools?

Your tooling. Do you have the kind of shop that's the envy of every woodworker or a modest collection of the usual hand and power tools?

Your skill level. Do you prefer simpler methods because you're starting out or are you always looking to challenge yourself and expand your skills?

The project. Is the piece you're making utilitarian or an opportunity to show off your best work?

In this book, we've included a wide variety of techniques to fit these needs.

To find your way around the book, you first need to ask yourself two questions: What result am I trying to achieve? What tools do I want to use to accomplish it?

In some cases, there are many ways and many tools that will accomplish the same result. In others, there are only one or two sensible ways to do it. In all cases, however, we've taken a practical approach; so you may not find your favorite exotic method for doing a particular process. We have included every reasonable method and then a few just to flex your woodworking muscles.

To organize the material, we've broken the subject down to two levels. "Parts" are major divisions of this class of techniques. "Sections" contain related techniques. Within sections, techniques and procedures that create a similar result are grouped together, usually organized from the most common way to do it to methods requiring specialized tools or a larger degree of skill. In some cases, the progression starts with the method requiring the most basic technology and then moves on to alternative methods using other common shop tools and finally to specialized tools.

The first thing you'll see in a part is a group of photos keyed to a page number. Think of this as an illustrated table of contents. Here you'll see a photo representing each section in that part, along with the page on which each section starts.

Each section begins with a similar "visual map," with photos that represent major groupings of techniques or individual techniques. Under each grouping is a list of the step-by-step essays that explain how to do the methods, including the pages on which they can be found.

Sections begin with an "overview," or brief introduction, to the methods described therein. Here's where you'll find important general information on this group of techniques, including any safety issues. You'll also read about specific tools needed for the operations that follow and how to build jigs or fixtures needed for them.

The step-by-step essays are the heart of this book. Here a group of photos represents the key steps in the process. The accompanying text describes the process and guides you through it, referring you back to the photos. Depending on how you learn best, either read the text first or look at the photos and drawings; but remember, they are meant to work together. In cases where there is an

The "VISUAL MAP" tells you where to locate the essay that details the operation you wish to do.

The "OVERVIEW" gives you important general information about the group of techniques, tells you how to build jigs and fixtures, and provides advice on tooling and safety.

alternative step, it's called out in the text and the visual material as a "variation."

For efficiency, we've cross-referenced redundant processes or steps described in another related process. You'll see yellow "cross-references" called out frequently in the overviews and step-by-step essays.

When you see this symbol, ⚠ make sure you read what follows. The importance of these safety warnings cannot be overemphasized. Always work safely and use safety devices, including eye and hearing protection. If you feel uncomfortable with a technique, don't do it, try another way.

At the back of the book is an index to help you find what you're looking for in a pinch. There's also list of further reading to help you brush up on how to use tools and keep them sharp, as well as some general references on design.

Finally, remember to use this book whenever you need to refresh your memory or to learn something new. It's been designed to be an essential reference to help you become a better woodworker. The only way it can do this is if you make it as familiar a workshop tool as your favorite bench chisels.

—The editors

"STEP-BY-STEP ESSAYS" contain photos, drawings, and instructions on how to do the technique.

"CROSS-REFERENCES" tell you where to find a related process or the detailed description of a process in another essay.

A "SECTION" groups related processes together.

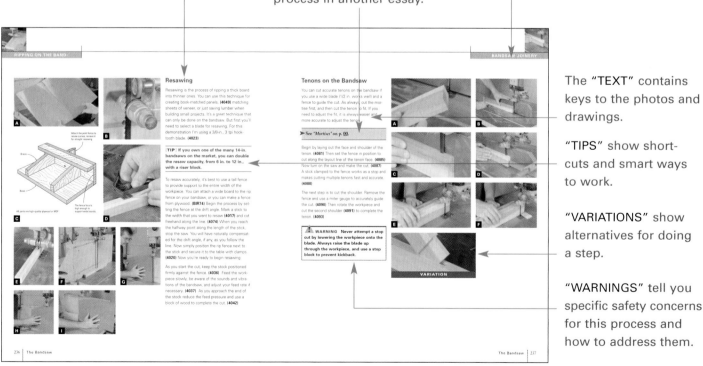

The "TEXT" contains keys to the photos and drawings.

"TIPS" show short-cuts and smart ways to work.

"VARIATIONS" show alternatives for doing a step.

"WARNINGS" tell you specific safety concerns for this process and how to address them.

PART ONE

Wood and the Shop

THE MATERIALS we select, the tools that we use, and even the space we work in combine to dramatically shape our woodworking. There's no right formula for how these ingredients come together because every woodworker's path is unique. So this first part of the book discusses wood as a material and the tools you'll need to start working it into something useful and attractive.

Many aspiring woodworkers dive right into accumulating an arsenal of tools and machines, only to find many gathering dust years later. To avoid this, I suggest you first learn a little about wood as a material, how a tree grows, how boards are harvested from trees, and why wood needs to be dry before you work with it. This knowledge, presented in Section 1, will pay valuable dividends throughout your life as a woodworker.

Section 2 offers advice for setting up shop, anchored by a solid workbench, and finding the right mix of quality hand tools and sturdy machines to do the kind of woodworking that interests you most.

Working with Wood

Understanding wood is essential to successful woodworking. Without knowing how to "read" grain direction, you may experience frustrating tearout while jointing or planing, spoiling a beautiful piece of lumber. Grain can also make a big difference in certain milling and shaping operations, such as making moldings. And working wood that's been dried insufficiently or too quickly can have unfortunate results, either while you're working with it or (perhaps worse) soon afterward. To help you get the most from the lumber you purchase, this section covers some of wood's most important characteristics, how to choose and store lumber, and how to find sources for eye-catching lumber at a reasonable price.

The Structure of Wood

In a living tree, the cell structure is much like a bundle of drinking straws transporting water—sap—up through the trunk and out to the leaves. Once the tree is cut down and sawn into planks, the orientation of the cells

forms what woodworkers refer to as grain. (Note: I'm discussing grain as the structural building block of wood. Grain is commonly used to refer to the visual characteristics of a board, but I prefer the word "figure" for this use.) There are two basic types of grain relevant to making things from wood—long grain and end grain. Long grain comprises the longitudinal surfaces of the cells—that's what you see on the faces and edges of a typical board. End grain is a cross section of the cells—what you see when you crosscut a board. The differences between the two have important effects on the working characteristics of wood.

When the long-grain surfaces of two boards are joined with glue, such as when gluing two boards edge-to-edge to make a wide panel, the completed joint is stronger than the surrounding wood. Put another way, if you try to break a glued-up panel along the glue line, the wood will break, perhaps near the glue joint, but the glue joint itself will hold. In contrast, when end

CROSS SECTION OF LOG

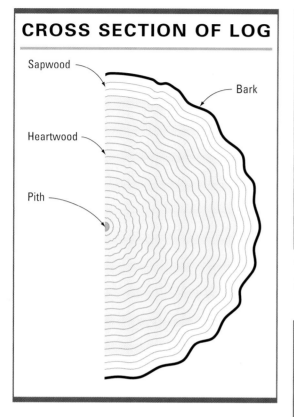

Sapwood

Bark

Heartwood

Pith

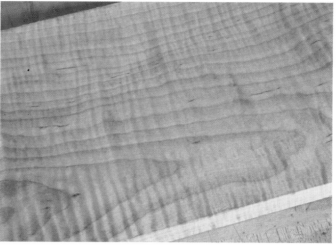

Tiger maple has distinctive stripes that run perpendicular to the primary figure of the wood.

TYPES OF GRAIN

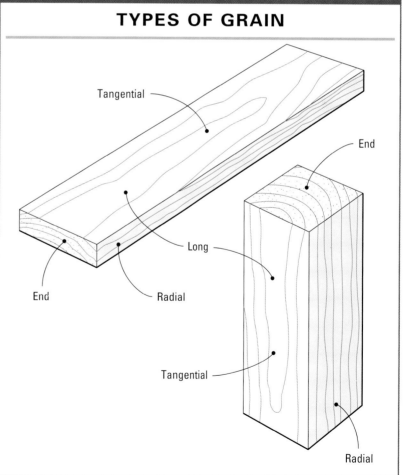

Tangential

End

Long

End

Radial

Tangential

Radial

grain is glued, such as in a right angle in drawer construction, the joint is weak and easily broken. Adding fasteners, such as nails or screws, may increase strength, but the fasteners may actually weaken the joint as they crush, break, and distort the wood's cell structure. The many joints used in woodworking were invented to solve this problem by creating contact between long-grain surfaces in both members.

Among the strongest, longest-lasting joints are those that interlock, such as the mortise-and-tenon and the dovetail, because they not only create long-grain gluing surfaces but form a mechanical connection as well. When precisely fit, interlocking joints stay assembled without glue. The test for a well-made mortise-and-tenon joint is a friction fit. Most woodworking glues don't fill

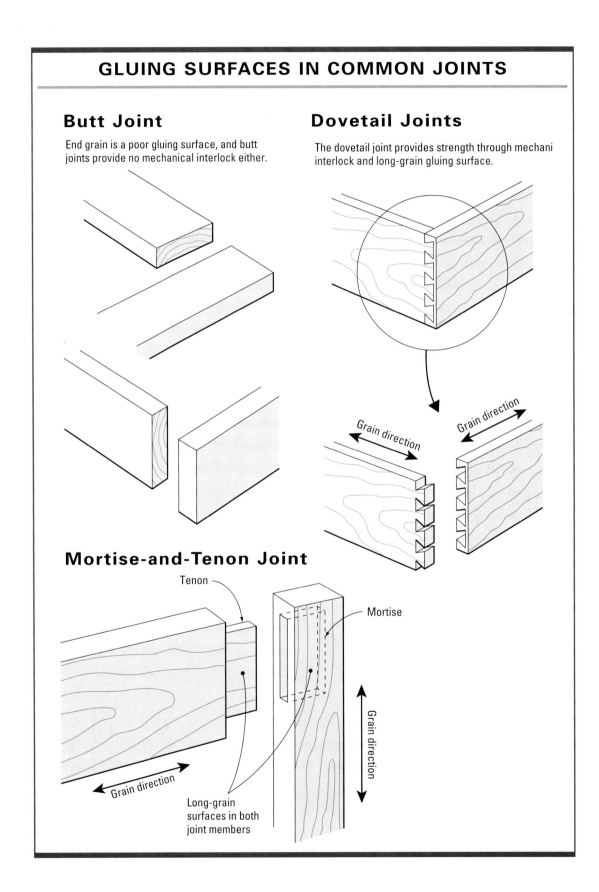

GLUING SURFACES IN COMMON JOINTS

Butt Joint

End grain is a poor gluing surface, and butt joints provide no mechanical interlock either.

Dovetail Joints

The dovetail joint provides strength through mechani interlock and long-grain gluing surface.

Grain direction

Grain direction

Mortise-and-Tenon Joint

Tenon

Mortise

Grain direction

Grain direction

Long-grain surfaces in both joint members

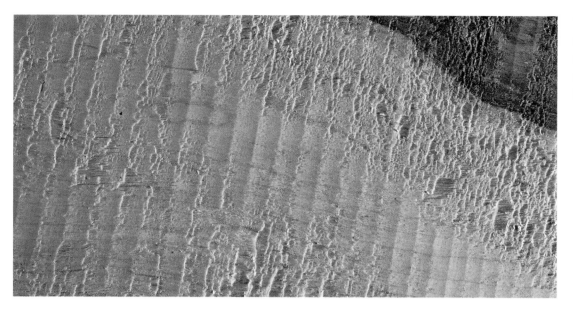

Tearout like this occurs when the stock is planed against the direction of the grain.

gaps between the mating parts. To make a good joint, glue must be uniformly applied to long-grain surfaces that contact fully and evenly when assembled.

The two basic types of grain, long grain and end grain, also work very differently with edge tools such as planes. Planes designed for end grain have a slightly lower cutting angle to help cleanly sever the tough ends of the cells. Planes for smoothing the long-grain surface of boards have a higher cutting angle to help prevent lifting and tearing the grain. Because grain seldom runs perfectly parallel to the surfaces of a board, planing requires that you follow the grain direction to limit tearout; this isn't always easy because sometimes the grain will change direction or seemingly go in several directions in the same board. Learning to read the grain takes practice but will prevent unsightly tearout when you're using a hand-plane or machines. (See p. 126, for more on planing with the grain.)

Grain plays a major role when you're using chisels, too. Chisels are designed for either chopping or paring. Chopping is severing the ends of the cells or grain. Because end grain is tougher and more dense than long grain, a chisel designed for chopping is ground with a steeper bevel and driven by the impact of a mallet. Paring chisels have a sharper, shallow bevel that enables the tools to remove long, slender shavings along the grain of the wood with minimal resistance.

From Logs to Boards

How the log was cut at the sawmill determines how the long grain meets the surface of the planks. This will affect the appearance of the board, its working properties, and its dimensional stability. All wood, even kiln-dried wood, expands and contracts with changes in the weather. More specifically, as the relative humidity rises, wood absorbs moisture vapor from the surrounding air and expands. During dry weather, typically the winter months, wood releases moisture vapor and contracts. The degree and the direction that it expands and contracts largely depends on how it was cut at the sawmill. Most logs

SAWING AROUND THE LOG

Standard commercial method that yields flatsawn boards

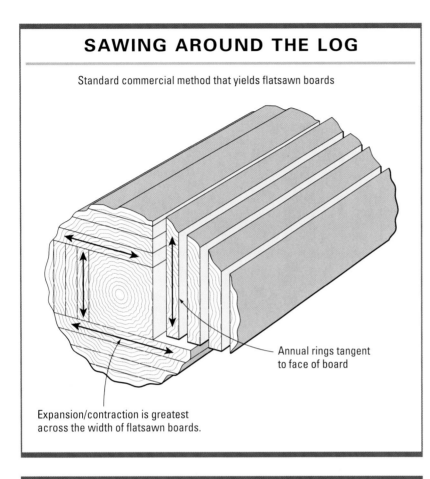

Annual rings tangent to face of board

Expansion/contraction is greatest across the width of flatsawn boards.

QUARTERSAWN LUMBER

Expansion/contraction is greater across the thickness than the width but marginal in both directions.

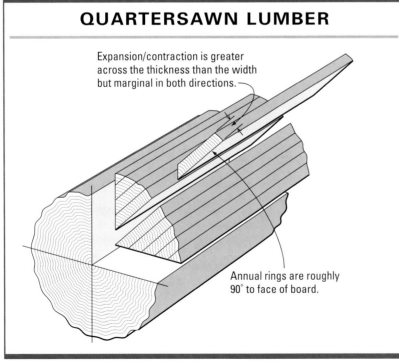

Annual rings are roughly 90° to face of board.

are sawn in a specific sequence of cuts so that the blade is always tangent to the annual growth rings. Referred to as "sawing around the log" in the lumber industry, this sawing method produces boards commonly called "flatsawn." It's efficient and produces the least waste. These boards are easy to spot; the surface displays the familiar "cathedral" pattern. Unfortunately, flatsawn boards react dramatically to changes in humidity and will cup, bow, and twist, even after being properly dried and milled. Most important, flatsawn boards expand and contract across their width to an appreciable degree, and that movement must be taken into account whenever you work with flatsawn lumber.

Radially sawn lumber, usually referred to as "quartersawn," looks and behaves quite differently from flatsawn lumber. In quartersawing, the blade is positioned radially, similar to the spokes on a wheel. In a quartersawn plank, the annual rings run in straight, parallel lines to the edges of the board. While this even, straight-grained pattern may seem boring, some species, such as oak, display a flashy "ray" figure when quartersawn.

In some species, like this white oak, quartersawn lumber displays a dramatic ray figure.

Quartersawn lumber is much more stable than flatsawn lumber. Because the growth rings of the tree run perpendicular to the face of the board (instead of at a tangent to the face) quartersawn lumber expands and contracts across its width far less than flatsawn. It also stays flatter after it's been planed. These are two good reasons that quartersawn spruce is used for the wide, thin soundboards on acoustic guitars. In fact, before the days of manmade sheet materials like plywood and particleboard, quartersawn lumber was used as a substrate for veneer.

Unfortunately, it's both time-consuming and wasteful to saw a log this way. A century ago, when forests seemed inexhaustible, quartersawn lumber was much more common. In that era, factories mass-produced furniture made of quartersawn oak, and many homes had quartersawn hardwood floors throughout the house. Because it's more stable and has less tendency to warp, quartersawn lumber wears more evenly, an important quality for flooring.

A third method for sawing logs yields boards that display both flatsawn and quartersawn figure. Sawing "through-and-through" creates wide boards with tangential figure in the center gradually shifting to straight radial figure along the edges. Many furniture makers, including me, prefer to have logs sawn this way for the broad, natural-appearing boards that this method produces. Also, the boards can be easily stacked for drying just as they come off the log. This makes it easy to match boards for grain and color when selecting lumber for a project. Because the boards are wider than those sawn with the two previous methods, you can often use one wide board in a door

➤ CROTCH GRAIN

Sometimes referred to as crotch figure, this flashy wood is a real eye-catcher. It's formed by the merging grain in the fork of a tree or the junction of a limb with the main trunk. To make the most of a crotch plank, it's often resawn into veneer. You can find crotch-wood planks at specialty lumber dealers.

Crotch wood comes from the fork in a tree trunk.

SAWING THROUGH-AND-THROUGH

This method yields wide boards with a natural appearance.

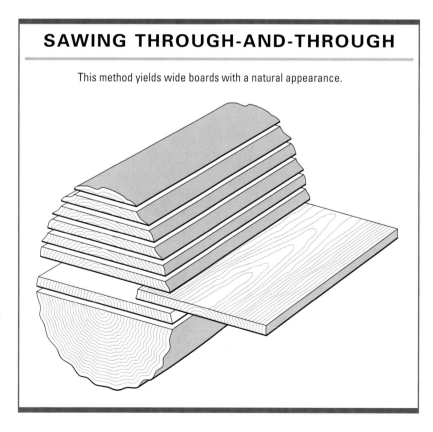

▶ AVOIDING SHORT-GRAIN PROBLEMS

As the name suggests, short grain is a condition where the normal strength of long grain is compromised and weak. It occurs most often when the grain doesn't run parallel to the edges of the stock. Sometimes that happens naturally, for example, in highly figured board, but it can also occur when a tight, bandsawn curve cuts across the grain. Either way, the result is a weakened area that is prone to breakage.

To avoid this dilemma:

• Select straight-grained stock for frames and other places where strength is critical.

• During layout, orient curved patterns so that the grain in the stock runs parallel to the length of the pattern.

• Avoid sharp 180-degree arcs that cut across one board. Instead, make arcs from two pieces of stock.

• Avoid bandsawing tight curves—laminating or bending are better methods. With either of these techniques the grain follows the curve, which adds tremendously to the strength.

Short grain is weak and prone to breakage.

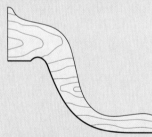

Orienting the grain this way results in short grain in the ankle area and may lead to breakage.

Orienting the grain this way avoids short grain at the point of greatest stress.

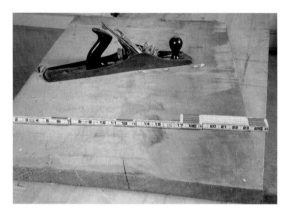

Wide boards are beautiful and dramatic, compared to several narrow boards glued together.

The center of the tree, known as the pith, is typically flawed and unstable and should be avoided when choosing boards for projects.

panel, desk lid, or small tabletop instead of gluing together two or more narrow boards.

For efficiency, logs at a commercial mill are sawn to create the highest grade of boards. When sawn through-and-through the widest boards, those nearest the center of the log will contain knots and other natural defects, which reduces their commercial value. However, for the small shop woodworker, it's easy to work around the knots or perhaps even include them as an interesting detail.

THE EFFECTS OF WOOD MOVEMENT

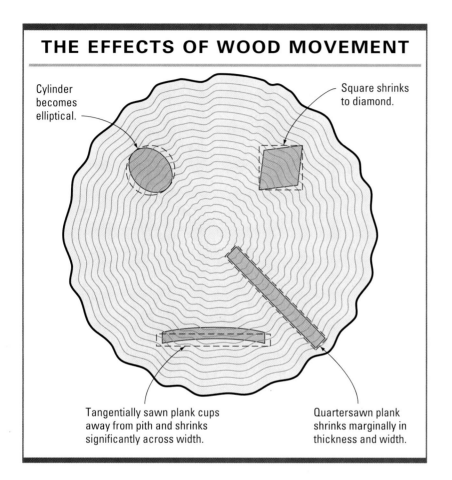

Cylinder becomes elliptical.

Square shrinks to diamond.

Tangentially sawn plank cups away from pith and shrinks significantly across width.

Quartersawn plank shrinks marginally in thickness and width.

Dealing with Wood Movement

Wood is hygroscopic—the technical name for the characteristic that causes wood to expand and contract with changes in humidity. It occurs even in kiln-dried wood and after applying finish. Fortunately, wood moves mostly across the grain, not lengthwise. (The movement along the length is so small that it can safely be ignored.) But because of the cross-grain movement, any joint that meets at right angles has the potential to crack, split, or otherwise fail as the two members move perpendicular to each other.

Here are some strategies for minimizing and managing the effects of wood movement:

- Use dry wood—6 to 8 percent moisture content for most areas of United States. You can purchase kiln-dried wood, or you can dry it yourself.
- Store the wood in a dry place—wood stored in a basement or unheated storage building will absorb moisture.
- Monitor the moisture content and relative humidity—you can purchase inexpensive meters to keep tabs on water in the air—and in your lumber.
- Construct for movement—design and construct joints that will allow the small but inevitable wood movement that will occur.

Begin with Dry Wood

Unless you're steam-bending a curved back for a chair or turning a greenwood bowl, your project should begin with dry wood. Just how dry should it be? Think about it this way: The wood must contain the correct level of moisture to place it in equilibrium with its environment. The heating and cooling systems in your home will dry the air. So the relative humidity indoors is usually lower than the air outside. As a rule of thumb, lumber for woodworking projects in most areas of the United States should have a moisture content of 6 to 8 percent. This will place the wood in balance with the 30 to 40 percent relative humidity in your home. Dry wood is preshrunk and prewarped, but lumber that contains too much moisture will shrink and warp after you've constructed that prized chest-of-drawers. So check the

moisture content of your lumber with a moisture meter before you start working it.

There are two common methods for drying lumber: kiln drying and air drying. Kiln drying is done by heating the wood in a closed container. The heat drives out the moisture quickly, usually over just a few weeks' time. I primarily use air-dried lumber that I dry myself. While a slower process than kiln drying, it has the advantage of allowing the wood to naturally reach equilibrium with the humidity of the surrounding air. Even kiln-dried lumber may be too "wet" if it's been stored for any length of time in an unheated or uncooled storage shed at the lumberyard. Lumber stored in an unheated space such as a barn will never drop below 12 to 15 percent moisture content. Instead of assuming that you can use kiln-dried wood without moisture problems, check it with a moisture meter. If necessary, allow it to acclimate to an environmentally controlled room, such as a heated or air-conditioned shop.

Storing Lumber

Regardless of whether you use kiln-dried or air-dried lumber, store it in a dry, stable environment. Obviously, if you haul a truckload of kiln-dried lumber through a rainstorm, it will not be dry once you get it to the shop, nor will lumber that is stored for any length of time in a damp basement or unheated garage. Before I begin a project, I store the lumber in my shop, which is centrally heated and cooled, and check the wood weekly with a moisture meter until it has reached equilibrium with the relative humidity in the shop. I stack it with sticks separating the layers (called "stickering") and

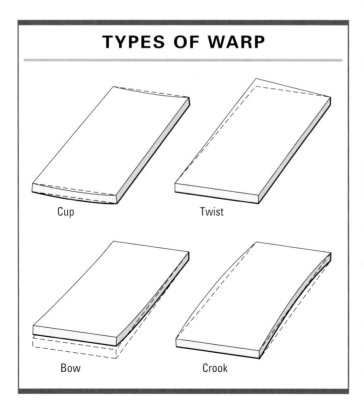

TYPES OF WARP

Cup

Twist

Bow

Crook

Positioning sticks evenly between boards promotes air circulation and prevents warping.

allow the air to circulate. If you don't have the luxury of an environmentally controlled shop, consider purchasing no more lumber than you can use within several weeks' time. Once you've flattened lumber and planed it to the final thickness, you can keep it from warping until you use it by wrapping it in plastic.

Plastic wrap will seal small boards or parts and prevent warping while you're waiting to use them.

Checking Moisture Content

One of the best investments you can make is in a small, handheld moisture meter. These devices measure the electrical conductivity in a board (water conducts electricity; wood does not) and convert it to moisture content, expressed as a percentage.

I prefer pin-type meters, though they're best suited to rough-sawn lumber. Pinless moisture meters are most accurate on lumber that's already planed. Be aware that the moisture content on the outside of the stock can fluctuate widely. In fact, I've seen an

For the most accurate moisture-level reading, cut a small amount from the end of a board and position the probes in the center of the end cut.

Pinless moisture meters work best on stock that's already planed.

A hygrometer will enable you to keep tabs on the relative humidity in your shop.

WOOD MOISTURE CONTENT

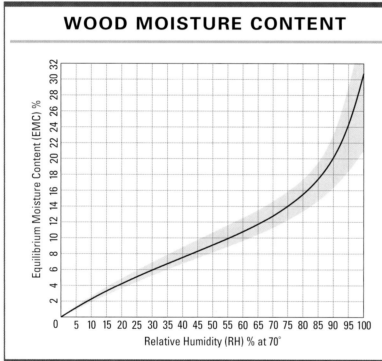

approaching storm raise the moisture content on the face of the stock in just a few hours. The most accurate reading comes from cutting a small portion from the stock and measuring the core.

Another good investment is a hygrometer. Mounted to the wall of your shop, this inexpensive box of electronics will allow you to

easily keep an eye on relative humidity. The graph in shows how relative humidity directly relates to moisture content. I keep a copy posted on the wall adjacent to the hygrometer. The relative humidity should be averaged in order to gauge the likely moisture content of the wood. For example, if you've got the wood for a project acclimating in the shop, record the relative humidity once each day over two weeks or more and take the average. Using the graph, you can determine the likely moisture content of the wood. If your shop is in the basement, you may find that the actual moisture content of the wood never gets below 12 percent. Sure, the wood has acclimated to the shop, and building your project may go smoothly. But when the finished project is moved inside the house,

► BUYING LUMBER

Hardwood lumber is expensive and is usually the single largest expense in any woodworking project. Fortunately, there are several ways to get good-quality wood and avoid reaching so deep into your pockets.

- Purchasing rough lumber can significantly cut the cost of your lumber bill. And "dressed" boards may be smooth, yet cupped and warped. Save yourself money and get flat, square lumber by purchasing rough lumber and milling it yourself. Nowadays, there are affordable small-to-medium-sized planers and jointers. Take a block plane along when shopping for rough lumber so you can skim the rough surface to reveal the board's figure.

- Buy from a hardwood lumber dealer rather than a general-purpose home center. While they're great places to buy plumbing supplies or paint, home centers charge top rates for those select and shrink-wrapped boards they conveniently stock.

- Buy No. 1 Common; this grade is one notch below the best ("select") and is often two-thirds to half the price. Of course, the boards will have knots and a few other defects, but you can usually plan your cuts around them.

- Purchase large quantities of 500 or 1000 board feet, and many lumberyards will give you a quantity discount. Consider splitting the cost with other woodworkers. Your local woodworking club may be a good source for other woodworkers who want to take advantage of group buying power.

- Look for small sawmills, those "little" guys who don't have the large overhead or the prices that go with it. These "sawyers" usually have a small, portable bandsaw mill that can take advantage of local trees felled for clearing land. Although some may have a kiln, most don't. So the lumber may still have to be dried further.

- Buy green lumber, which is my personal favorite for preserving the lumber budget. Often green lumber (still wet with sap) can be had for one-fourth to one-half the going price for dry lumber.

Lower grades of wood, including sapwood, have more defects, but you can often cut the defects out and use the remaining wood.

where the relative humidity is far lower than in the shop, the wood will shrink. That could result in an array of problems, from cracked panels and tabletops to overly loose-fitting drawers. The remedy is to lower the relative humidity in the shop, and a simple step toward that goal is to use a dehumidifier.

Drying Your Own Lumber

You can buy undried lumber from small sawmills or local loggers, but be prepared to accept the boards as they come. Sometimes you can cut a deal before the log is even sawn and have it sawn the way you want. As I mentioned earlier, my favorite sawing method is through-and-through. You'll get much wider boards, but be aware that you'll also get some lower-grade boards that con-

tain knots and pith. As an advantage, you'll keep the sawmill operator on your side because sawing through-and-through is by far the easiest and least time-consuming method.

Once you've acquired recently cut boards, you'll need to dry them before use. You can stack the lumber outdoors, but the best place to dry lumber is inside a barn or shed. This will avoid the repeated exposure to rain and sun that can cause the boards to split and warp excessively. Otherwise, try to locate the stack under a shade tree. If the stack is outdoors, you can use inexpensive construction-grade plywood to shed the rain. Weight the plywood down to prevent it from blowing away in a windstorm. Avoid the temptation to use plastic as a cover. It won't allow

▶ MAKING STICKERS

"Stickers" are just sticks of wood used to separate the layers of green lumber. This allows air circulation, promotes drying, and helps avoid mildew stains on the stock. I usually collect drying sticks as I'm woodworking by saving any long, narrow offcuts. Avoid using green lumber left over from sawing the logs. The moisture can often cause staining that will be difficult to remove. It also helps if you saw the sticks to a uniform size, ¾ in. by ¾ in. or 1 in. by 1 in., so that the stack rises uniformly.

STACKING GREEN LUMBER

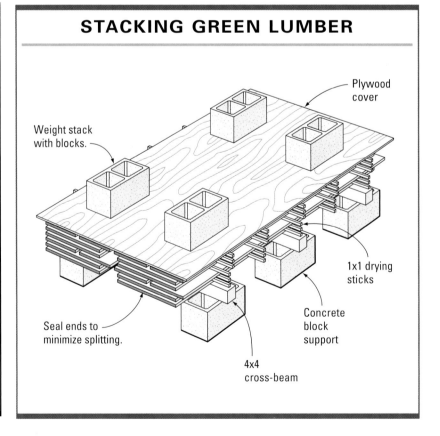

Plywood cover

Weight stack with blocks.

1x1 drying sticks

Concrete block support

Seal ends to minimize splitting.

4x4 cross-beam

► THE IMPORTANCE OF SEALING THE ENDS

In life, the cells of the tree were an efficient plumbing system for sap. Once cut, they provide a conduit for moisture to escape. As green wood dries, the cut ends will release moisture too rapidly, causing the ends to shrink and split. Sealing the ends minimizes and in some cases eliminates splitting.

Cut the drying checks from the end of the lumber before sizing it.

enough air circulation, and mold may develop. Be sure to choose a dry location outside; otherwise the lumber will take longer to dry, and mold can develop in the process.

To minimize warping, give boards good support. I use concrete blocks for a foundation and 4-by-4 cross members. Space the blocks 3 ft. to 4 ft. apart, depending on the lumber thickness; thin lumber such as 4/4 and 5/4 has a greater tendency to sag under its own weight. As you stack the boards, leave at least 1 in. of space between them for air circulation. Use "stickers" (see the sidebar on the facing page) to provide space between layers. I usually position the sticks about 12 in. apart, further if the lumber is thick. Before the stack grows beyond a comfortable height, stop. Tall stacks may topple over in the wind, and lifting heavy, wet lumber up beyond shoulder height is a lot of work.

Sealing the ends of the boards can be done anytime in the process, but sealing the logs before sawing them into planks is more efficient. Apply green lumber sealer (latex paint works too) with a paintbrush or roller. As the wood dries, monitor the moisture. One year of drying for each inch of thickness is the old rule of thumb, so you'll have to be patient. Once the lumber has reached equilibrium with the outdoors, it is time to bring it indoors for further drying or take it to a local kiln for kiln drying.

Outfitting the Shop

BEFORE SPENDING A LOT of money on tools and equipment, consider the area of woodworking you would like to pursue. For example, if you're primarily interested in constructing kitchen cabinets and built-ins, you'll probably need more power tools and the space to operate them. However, if you enjoy carving, the bandsaw and a lathe are perhaps the only machines you'll need.

No matter what, you should obtain a sturdy workbench, along with a vise and clamps for holding work. When considering hand tools for purchase, buy the best you can afford; cheap hand tools are no bargain and tend to lead only to frustration. In the following section, I'll guide you through many of the tools and machines that you should consider for your shop.

Holding the Work

While sawing, planing, chiseling, and routing, you'll need a sturdy workbench to hold the work. Your workbench need not be fancy, but it should be strong and heavy to resist pounding and racking. Additionally, the best workbench is one that fits you, so I would avoid purchasing this essential tool unless you can modify the height to suit your height and needs. Consider buying the materials and making it yourself to fit your height and the space available in your shop. To hold the workpiece, you'll need to outfit your bench with a vise. Purchase the largest vise you can afford and mount it on one corner of your bench so you'll have the greatest access to the workpiece.

Good lighting is important, too. Locate your bench near a window to take advantage of the natural light. And use plenty of electrical lighting as well; fluorescent lights work well overhead while incandescent fixtures provide close-up lighting.

A solid workbench is an essential tool for any wood-working shop.

Handscrews apply broad clamping pressure, even on nonparallel surfaces.

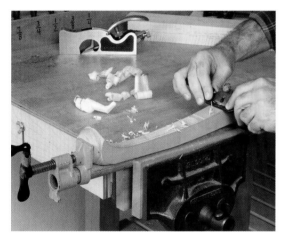

A strong vise completes the workbench. A pipe clamp locked in a vise is a great way to hold irregular work, such as this cabriole leg.

Clamps play an important role in any shop; they hold glued-up assemblies while the glue sets, and they secure work to the benchtop. The old favorites, such as pipe clamps and handscrew clamps, are versatile performers and a great choice for your first clamp purchases. Later on, as your wood-working skills and shop expand, you can purchase more clamps as the specific need arises.

Pipe clamps are inexpensive and useful for a wide array of assembly tasks.

A modest tool kit consists of a few planes, a couple of chisels, and a dovetail saw.

Chisels are available in a variety of shapes and sizes to suit the job at hand.

Carving chisels and gouges let you create fine details that can only be done with handwork.

Edge Tools

Edge tools include chisels, planes, handsaws, scrapers, and gouges for turning and carving. This collection of hand tools shapes, cuts, smoothes, and otherwise creates the details in the furniture that machines simply cannot.

Chisels are available in various shapes, widths, and lengths. The most common, bench chisels, are used to cut joinery and fit hinges and other hardware. Carving and turning gouges create shapes for adding decorative embellishments to your work. For the greatest control with any chisel or gouge, it is important to keep it sharp.

Planes are the workhorses among hand tools and are used for smoothing, shaping, and fitting joints. A finely tuned bench plane will create a smoother surface than any sander. Bench planes can also be used for flattening and squaring stock that is too large for your jointer. For one-handed trimming and fitting, especially of small parts, include a block plane in your tool kit; low-angle block planes are ideal for trimming end grain. Shoulder planes are finely tuned

A bench plane, shoulder plane, and block plane form the foundation of a versatile plane kit.

The block plane excels at light trimming.

Molding planes create profiles with a hand-made look.

precision tools designed for careful fitting of joints. They feature precisely ground soles and irons for removing feathery-thin shavings.

Wooden molding planes feature a profiled sole and matching iron that create a contoured surface. Many molding planes, such as hollows and rounds, were once produced in mass quantity and are still available today. These beautiful old planes are still useful for shaping large moldings beyond the capabilities of your router table.

➤ HOW EDGE TOOLS CUT

You'll be able to sharpen, tune, and use your edge tools most effectively, and you'll get more enjoyment from them, if you understand the dynamics of the cutting edge.

An edge is formed by two surfaces that intersect; the angle and sharpness of the intersection directly affects how well the edge cuts. A sharply honed edge will slice the wood and create a thin, delicate shaving while a dull edge works more like a thick wedge to splinter the wood.

A low-bevel angle, 20 degrees, for example, will slice more cleanly and with less resistance than a higher angle of 30 degrees. Of course, there are trade-offs, too. A lower angle weakens the edge, which makes it prone to fracturing. (This is why you should avoid striking paring chisels with a mallet.) Also, a lower-bevel angle is not as effective at breaking and curling the shaving, which can lead to tearout in difficult grain (such as curly maple).

A higher-bevel angle, such as 30 degrees, is stronger (which is why it works well for mortising chisels) but also has a greater cutting resistance. An angle of 25 degrees is a compromise that works well for most chisels and plane irons.

A plane is essentially a souped-up chisel. As the cutting edge lifts the shaving, the sole holds it down while the cap iron breaks and curls it. This is why it is essential to set the cap iron close to the edge of the blade and close the mouth of the plane as tightly as possible.

The spokeshave is a small plane ideal for smoothing curves.

A bench hook is a companion to the backsaw; it cradles the workpiece perfectly as the cut is made.

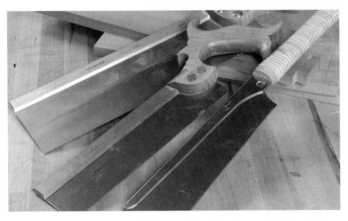

All backsaws have a thickened back edge, which stiffens the blade for cutting fine joinery.

When smoothing bandsawn surfaces, I reach for a spokeshave. This small plane has a short sole and handles on each side, which makes it the ideal tool for smoothing curved stock.

Handsaws are for cutting fine joinery and cutting away large portions of stock before planes refine the surface. Dovetail joinery, for example, is cut with a small backsaw, known as a dovetail saw. Backsaws work best for joinery because they cut a fine kerf, and the back is reinforced with a brass or steel spine. Coping saws have a narrow blade designed for cutting curves and scrollwork.

Saws are designated by the teeth, either rip or crosscut. Rip teeth are designed for cutting parallel to the grain, while crosscut teeth are shaped like tiny knives to cleanly sever the wood when cutting across the grain.

Western saws cut on the push stroke while Japanese saws cut on the pull stroke. Their smooth yet aggressive cutting has

A rip-tooth saw works best for dovetails.

Scratch stock are quiet tools for scraping simple profiles.

The cabinet scraper will smooth even the most difficult woods.

made Japanese saws a favorite among many woodworkers.

You can cut your sanding time more than in half by using a scraper. This tool smoothes wood with a small burr, and when sharp, it produces shavings like a plane yet it doesn't tear the wood like a plane sometimes can. A scratch stock is a profiled scraper used for shaping small moldings. This is another tool that will shape profiles that are difficult or impossible to shape with a router.

Measuring and Marking Tools

The old adage "measure twice, cut once" is still good advice. Most every woodworking project begins with careful measuring and marking, otherwise known as layout. Layout tools consist of rules, tapes, squares, dividers, and marking knives, to name a few.

The 6-ft.-folding-wood rule is still my favorite tool for measuring. This vintage-style rule folds compactly to slip easily into a pocket. Steel tapes quickly retract onto a spool for storage and are useful for measuring long lengths of rough lumber.

Accurate layout work begins with an assortment of essential layout tools.

The folding rule, though less common than the measuring tape, is still an accurate and convenient measuring tool.

A machinist-quality combination square should be among the first layout tool purchases.

The bevel gauge is used to lay out, copy, and transfer angles.

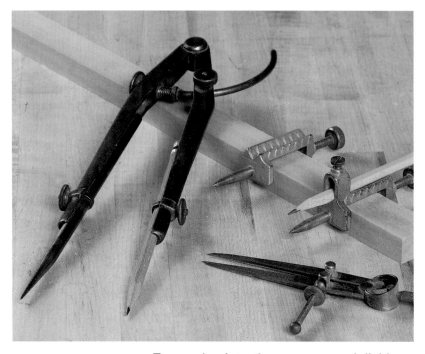

Trammel points, the compass, and dividers are useful for layout of circles and arcs.

A reliable square is a tool that no shop should be without. My favorite is a 12-in. combination square. This multipurpose tool serves as an inside square, outside square, 45-degree square, depth gauge, and straight-edge. Get the machinist-quality square; those sold at home centers lack the quality and accuracy for fine work.

The sliding-bevel gauge is used for laying out and checking angles other than 90 degrees. The steel blade pivots and locks in place at virtually any angle.

Dividers, trammel points, and compasses all mark a space between two points. Dividers are used to transfer measurements and step off linear dimensions. The compass is similar to a divider, except a pencil is substituted for one of the points. The compass is the tool of choice for drawing small arcs and circles. Trammels are used for drawing large circles, such as a tabletop, that are beyond the reach of a compass. Trammels come in a pair and

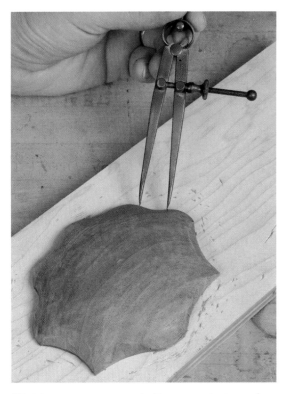

Dividers measure and allow you to transfer points on irregular shapes, such as carvings.

Trammels are used to lay out circles and ellipses.

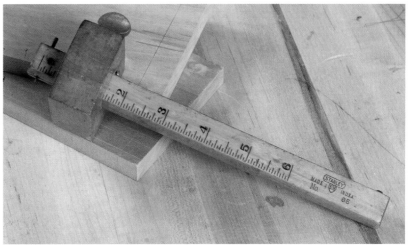

A marking gauge inscribes the wood with a tiny knife.

A marking knife is an essential layout tool.

clamp to a stick of any length. By adding a third trammel to the stick, you can draw an ellipse.

A marking gauge and layout knife belong in every tool kit. These tools incise the wood, unlike a pencil, and create a sharp guideline for chiseling and sawing. The best marking gauges have a graduated beam that makes it easier to set the tool for a precise measurement. Although you can purchase expensive layout knives with rosewood handles, an X-Acto® knife is inexpensive, and the thin blade reaches into areas that are inaccessible to larger knives. And when the blade dulls, you can toss it out and replace it with a new one.

No shop should be without an assortment of hammers and mallets.

A dead-blow mallet (far right) provides a controlled, non-marring impact for assembly.

Hammers and Striking Tools

Hammers, mallets, and other striking tools are used to deliver a precise impact. Hammers are used to drive nails and brads and feature a crowned, or convex, face that helps prevent marring of the wood surface.

Dead-blow mallets feature a shot-filled head that eliminates bouncing upon impact. For assembling joints and casework, a dead-blow mallet is used for gently tapping parts into alignment.

Carving mallets are used to direct controlled force to a chisel or carving gouge. Some are made of dense tropical hardwood, while others feature a urethane head fitted to a wood handle. When selecting hammers and mallets for woodworking, choose the lighter ones, typically those that weigh under a pound. Greater force just simply isn't needed for most woodworking, and heavy mallets are both tiring and awkward to use.

After you've acquired a solid mix of hand tools, keep them organized and within easy reach. Small tools, such as chisels, squares, and files, store easily in a rack. The rack can be wall-mounted or attached to the back of your bench.

Carving mallets come in a variety of sizes for delivering a precise blow to gouges and chisels.

➤ BUYING OLD HAND TOOLS

It's difficult to produce high-quality work with inferior tools. But if you're on a budget (as most of us are), one way to extend your woodworking dollars is to shop for high-quality old tools. A trip to a tool auction or flea market will yield planes, marking gauges, and bevels by Stanley®. You'll also find squares, calipers, and dividers by familiar names like Starrett® and Lufkin®. Best of all, most of these tools can be added to your kit for a fraction of the cost of a new tool. Tools that show signs of use or need extensive cleaning and reworking are often passed up by collectors and can be a bargain for the woodworker. But there are things to avoid when buying old tools, too. Here's a list:

- *Tools with missing parts*. Avoid tools with broken, bent, or missing parts. Most parts are not standard hardware store items, and finding a replacement can be difficult or impossible.

- *Planes with cracked castings.* A plane with a cracked sole, even one with a good repair, may not ever function properly again. Besides, there are plenty of intact planes available that need only light cleaning and tuning.

- *Heavy rust.* Most old tools have at least some rust. But those that have spent years in a wet basement or old barn may be encrusted with heavy rust. Although you can clean it off, heavy rust will leave the surface pitted. If it's a chisel or other edge tool, the rust pits create a void at the tool's cutting edge.

Old hand tools are often of better quality than commonly available new tools.

A storage cabinet will organize and protect your hand tools.

A storage rack keeps chisels and gouges sharp and close at hand.

THE IMPORTANCE OF SHARP TOOLS

Sharp tools are a pleasure to use, as they slice thin shavings and leave the surface of the wood glistening. Sharpness gives you much better control of the tool and enables you to do your finest, most precise work. Sharp tools slice and shear the wood cleanly while dull tools crush and tear.

When I'm teaching woodworking classes, two things often surprise students—how quickly an edge dulls and yet how quickly it can be brought back to sharpness. Sure, stopping to sharpen a tool interrupts the work flow. But once you've learned the steps you'll soon be back to planing, dovetailing, and carving. You'll also have mastery of the tool, and your work will dramatically improve.

For larger tools, such as planes and saws, a cabinet works well for storage. I avoid storing tools in drawers, especially those tools I use often. Drawers easily become cluttered, and they hide their contents from view. So finding the tool you need can turn into a busy search. In contrast, cabinets and racks keep tools organized, sharp, and close at hand.

Sharpening Equipment

Although it's best to send router bits and circular saw blades out for sharpening, you'll want to sharpen your own hand tools. The steel in hand tools dulls quickly compared to the carbide on bits and saw blades, and so it isn't practical to send these tools out for sharpening. To keep your edge tools in top performance, you'll need a grinder and a set

To keep edge tools sharp, you'll need a set of sharpening stones.

Portable planers are affordable alternatives to a stationary machine.

of benchstones. Grinders feature a coarse stone wheel and rests or supports to hold the tool at the proper angle. Grinders are powerful tools that you can use to quickly restore the bevel of a chisel or plane iron. Afterwards, benchstones are used to hone the edge, which sharpens it further.

Portable Power Tools

Portable power tools allow you to take the tool to the workpiece. Sometimes the work is too heavy or awkward to maneuver through a stationary power tool. That's when portable tools, such as drills and jigsaws, take over. Some portable power tools, such as routers and miter saws, can be mounted to a table or stand and used as stationary equipment. Yet they cost much less and take up less space in your shop than large, heavy, stationary equipment.

These days, many portable power tools are cordless. They use powerful batteries that can be quickly recharged. This allows you to work without dragging an extension cord around the room.

The cordless drill has quickly become a shop favorite.

A jigsaw will cut curves on work too large to maneuver at the bandsaw.

The plate joiner cuts fast joints for simple cabinet construction.

With their narrow reciprocating blade, jigsaws excel at cutting curves. Although it won't replace the bandsaw for resawing, the jigsaw is a good option for cutting interior cuts or cutting curves on a workpiece too large to handle at the bandsaw.

Although biscuit joiners and sanders are not typically used for the finest work, these tools are great for quickly joining and sanding shop and kitchen cabinets and other less demanding jobs. Biscuit joiners use a small-diameter saw blade to cut a circular slot. The slot is cut in two mating pieces and joined with a compressed-wood plate, or "biscuit."

Portable sanders can efficiently level joints and smooth surfaces. However, these tools cut very aggressively and are not known for creating flat surfaces, so their best use is on inexpensive cabinetry. To smooth your finest work, reach for your bench plane and scraper.

Power sanders don't leave the surfaces flat, so avoid using them on your finest work.

The router is one of the most useful tools you can own.

The router table has replaced the shaper in many small shops.

Over the past few years, routers have dramatically changed woodworking. Once tools for shaping the edge of a tabletop or a small molding, routers can now cut joints, mortise for hinges and locks, and even shape the edges of raised panels. The transformation is due to the large selection of router bits, jigs, tables, and accessories that have been developed for this universal tool.

Although technically portable tools, routers often get the most use mounted in a table. This in effect creates a mini-shaper. Like shapers, router tables have fences and miter gauges to support and guide the workpiece. Best of all, mounting your router in a table allows you to use large bits that would be unsafe otherwise.

Fixed-base routers feature a motor that mounts in the router base and locks in place. Plunge routers are designed so that the motor and bit can be lowered into the workpiece while running. Plunge routers are most useful for cutting mortises and other types of joinery, using a jig to guide the cut.

Miter Saw

The popularity of the electric miter saw (sometimes called a chopsaw) has almost led to the extinction of the radial-arm saw. It's a portable tool, found on just about any job

▶ USING A MITER SAW

Undoubtedly the most useful purpose of a miter saw is for cutting miters on moldings and trim. Equipped with a sharp crosscut blade, most miter saws will cut an extremely smooth miter that needs no further handwork before assembly.

When attaching molding to a cabinet, such as the small, dovetailed box in these photos, I begin by planing the joints flush at the intersections. Position the plane at a 45-degree angle to avoid tearing out either of the joining pieces (A). Also, check the surfaces for trueness (B). As you can imagine, it's much easier to apply molding to a straight surface.

Miter one end of the front molding first (C). Then position the molding on the cabinet (D) and mark the miter at the opposite end (E). Now attach the front molding first (F). Then apply glue (G) and attach the side returns (H).

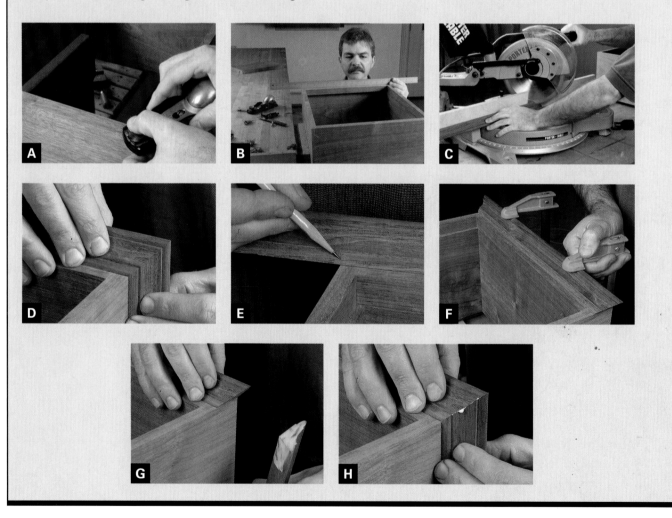

site where clean crosscuts or miters are needed. Mounted on a stand and upgraded with a high-quality blade, the miter saw functions more like a stationary power tool in most woodworking shops. It excels at making accurate, repetitive crosscuts and is far more convenient than the table saw for crosscutting long stock like moldings. The head of a sliding miter saw is mounted on rails that allow it to slide outward and cut wide stock beyond the capacity of standard miter saws. The head of a compound miter saw tilts for cutting crown molding and other work with compound angles.

When shopping for a miter saw, stick to the 10-in. models. Twelve-in.-diameter blades don't cut as accurately because the teeth are further from the support of the arbor. Besides, a 10-in.-diameter saw is large enough for most work.

Stationary Machines

Large machines have the heft and weight to dampen vibration. Generally speaking, they hold settings longer and produce smoother cuts.

Table Saw

Most people start out woodworking with a table saw. This universal machine can rip and crosscut stock to size; cut tenons, grooves and rabbets, among other joints; and even shape coves for moldings.

The cabinet-model table saw weighs in at a quarter-ton and features a cast-iron top supported by a heavy-gauge steel cabinet. The less expensive contractor-style saws reduce cost by using less iron and more sheet metal as well as an open stand. Either way, equip your saw with a guard and splitter and a high-quality saw blade.

A table saw is usually the first power-tool purchase. In addition to doing ripping and crosscutting, it's used for finished joinery and even shaping operations.

The jointer will true the faces and edges of stock.

Jointers and Planers

Jointers and planers work as a team to flatten and square rough stock and plane it to thickness. Fortunately, as the popularity of woodworking has risen, the price of these expensive machines has come down. If you can, avoid the 6-in. jointers and purchase an 8-in. or even a 12-in. machine. To create fine work, you'll want to flatten boards on the jointer before planing them to thickness. A 6-in. jointer is simply too small, and you'll often be reaching for a bench plane to flatten rough stock.

After the jointer flattens a face of the board, the planer mills stock to final thickness. Twelve-in. to 15-in. planers have become quite affordable, and this size is ideal for most small shops.

The planer is a workhorse that mills stock to size.

➤ COMBINE MACHINES AND HAND TOOLS TO BEST ADVANTAGE

Hand tools are a pleasure to use, but I don't want to ever handplane a whole pile of lumber for a large project. A power jointer and planer are the best tools for that job. For smaller jobs, and if you don't have a jointer yet, you might flatten the first face of all your stock with a handplane, but power plane the second face with a relatively affordable portable planer. I use hand tools for creating the fine details that machines can't create.

For example, you can flatten boards with a long bench plane and winding sticks (see p. 146 in planing section). However, you may soon become weary of this labor-intensive chore. Instead, consider purchasing a larger jointer. A 12-in. machine will handle 90 percent of the lumber that comes through your shop. After flattening the face on the jointer to remove cup or twist, use the power planer to smooth the opposite face.

The bottom line is this: Savvy woodworkers use both power tools and hand tools. Power tools can efficiently handle the labor-intensive chores. This will give you more time to enjoy using hand tools to cut dovetails, fit drawers, and carve, as well as creating other fine details that can't be produced with machines.

The Bandsaw

Like the table saw, the bandsaw is another universal machine that will surprise you with its versatility. The bandsaw is the tool of choice for sawing curves—all types of curves from broad and sweeping to tight and scrolling. And the bandsaw is also the only machine that will resaw, or rip, a thick board into thinner ones. Resawing is a great way to save money when making small projects that require thin boards. This technique can also be used to saw your own veneer from a figured plank. But that's not all. The bandsaw will also cut surprisingly accurate joinery, such as dovetails and tenons.

The 14-in. bandsaw is a widely popular size and will handle most any sawing task, especially when equipped with a riser block, which increases the resaw capacity from 6 in. to 12 in.

In addition to cutting curves, a good bandsaw can resaw boards and rip stock to rough width more safely than the table saw.

A drill press has far more power and precision than any hand drill.

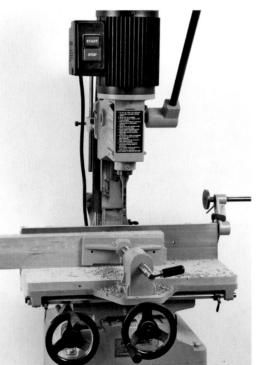

A hollow-chisel mortising machine creates precise mortises quickly and accurately.

Drill Press and Mortiser

The drill press will do what's nearly impossible with a portable drill—cut perfectly perpendicular holes. When I need a large-diameter, accurately bored hole, I turn to the drill press. Drill presses are inexpensive, and you can stretch your budget (and shop space) further with a benchtop model. If you build a dedicated stand for it, you can optimize the storage space that's wasted on a floor-model drill press.

Mortise machines are beefy little specialized drill presses. The mortise machine uses a drill enclosed within a square, hollow chisel to cut mortises. These machines will cut deep, accurate mortises quickly and efficiently. Benchtop models are inexpensive and easily stowed away inside a cabinet or in a corner when the job is finished.

The Shaper

The shaper is a powerful workhorse for shaping a kitchen full of paneled doors or miles of moldings. A shaper functions like a table-mounted router but with greater accuracy and lots more horsepower. However, this is a machine that you can postpone purchasing for a while. Later on, after you've gained plenty of experience with a router table, you'll be able to better determine whether you need a shaper.

The Lathe

The lathe spins stock while it is shaped with special gouges and chisels. You can use a lathe to turn a bowl or a decorative table leg.

Lathes vary tremendously in their size. The most common size is a 12-in. lathe with 36 in. between the head and tailstock. Before purchasing this specialized machine,

The shaper is a powerful machine that requires skill to operate effectively. Here it's shown with a power-feed attachment that ensures both accuracy and safety.

➤ HORSEPOWER

Consider motor horsepower when shopping for machinery. When you're making a heavy cut or during extended periods of use, motors can overheat and temporarily shut down. This can sometimes be avoided by keeping blades and cutters sharp and using a slower feed rate. But underpowered machines are no bargain. Skimping on horsepower can limit the potential use for a machine. It's important to compare horsepower when shopping for machines. Also keep in mind that when the cutter size increases (for example, from a 6-in. to an 8-in. jointer), the motor horsepower should increase as well. As you acquire woodworking machines, you may find it necessary to hire a licensed electrician to add designated electrical outlets to your shop—or perhaps even a larger service panel.

you may want to take a short course in turning to gain a few skills and determine your level of interest.

Combination Machines

As the name implies, combination machines combine the features of several machines on one piece of equipment. Most have a table saw, jointer, and planer, and perhaps a shaper. When you turn knobs and release levers, the machine makes a quick metamorphosis from one tool to another.

Although you may find it tedious to continually adapt the machine from one function to another, combination machines can be a great option for some woodworkers—specially those with limited shop space.

Acquiring Tools

If you're new to woodworking, you may find the process of purchasing so many tools exciting or overwhelming. Oftentimes woodworking catalogs promote almost every tool within the pages as one that you "must have." Also, too often jigs and gadgets are promoted as a way to avoid acquiring the skills needed for using basic tools. Just be aware that some of the finest furniture and woodwork ever produced was crafted before the age of power tools and dovetail jigs. It's tough to measure the enjoyment and satisfaction that come with using basic hand tools.

If you're excited about woodworking but are unsure of where to begin, a good place to start is with a few essential hand tools. A couple of planes, a few layout tools, a dovetail saw, and a set of bench chisels are a great foundation for any collection of tools.

Buying Machines

Fortunately, there are a number of companies producing affordable woodworking machines for the small shop. However, there are still some to avoid. It's best to visit a woodworking store or tool show and examine a machine before making a purchase. Fit, finish, and balance of moving parts have the most importance. It's difficult to do accurate work on a machine that vibrates excessively. Bandsaws are especially prone to vibration and require precise balancing of pulleys, wheels, and even the motor.

Blades, Bits, and Cutters

The cutting tool is the heart of any machine. A great blade will significantly boost the performance of an average saw, yet the finest machine will disappoint you when equipped with poor-quality tooling. High-quality blades, bits, and cutters are well machined and balanced, which insures smooth cutting.

Every power tool will perform better with quality bits, blades, and cutters.

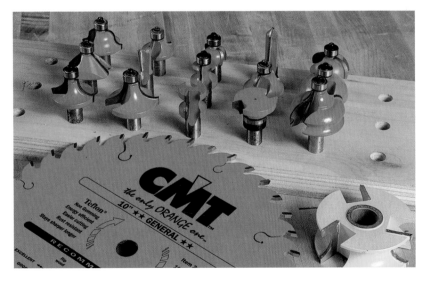

The best carbide tips are thick, which allows for repeated sharpening, and made from fine-grain carbide.

The edges of blades and bits can be carbon steel, high-speed steel, or carbide. Many bandsaw blades are made from carbon steel, the same type found on edge tools, such as plane irons and chisels. Although carbon steel takes a fine edge, it isn't as heat resistant as high-speed steel and carbide.

High-speed steel can be sharpened to a keen edge and is heat resistant—good qualities for tooling. In fact, many turning gouges are made from high-speed steel. However, high-speed steel will not wear as long as carbide. And high-speed-steel dulls quickly when used on man-made sheet stock, such as particleboard.

Carbide is a hard powder metal that is bonded and brazed to the tip of blades and cutters. Because of the extreme hardness of carbide, it is extremely long-wearing.

Be aware that many of the factors that are involved in producing high-quality tooling are difficult or impossible to physically examine. The quality is in the cut. One way to evaluate tooling is to read the reviews and comparisons in magazines.

Dust Collection

You can dramatically improve the working conditions in your shop by adding dust collection. Fortunately, the cost of dust-collection units has come down as the popularity of woodworking has increased. A portable unit on casters can easily be rolled around the shop and used where needed. As you add equipment, you may want to consider using a central dust collector. A central unit is powerful and convenient, and you'll

► WORKING SAFELY

Woodworking is inherently dangerous. Using tools improperly can lead to a serious injury or perhaps even death. Don't attempt to try the techniques in this book until you are sure that they are safe for you in your own shop.

Here's an additional set of guidelines that I follow in my own shop:

- Always read and follow the manufacturer's safety guidelines that come with any tool.

- Use guards that come with machines.

- Keep blades, bits, and cutters sharp.

- Use safety devices, such as push sticks and push blocks, to distance your hands from the cutters.

- Always wear eye and hearing protection.

- Don't force tools; if the tool isn't working properly, STOP!

This compact, two-stage dust collector will handle almost any machine in your shop.

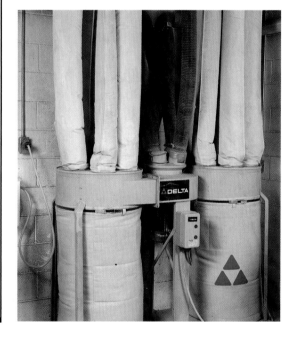

A large, central dust collector can handle several machines at once.

no longer have to push the portable unit around to each machine.

No dust collector traps 100 percent of the dust at the source. Here's where the ambient air cleaner takes over. Using a fan and a furnace-type filter, an ambient air cleaner traps the extremely fine dust that normally remains suspended in the air for several hours (giving you plenty of time to inhale it). Most ambient air cleaners have three speeds,

a timer for automatic shut-off, and a remote control. With an ambient air cleaner mounted against the ceiling, your shop will stay cleaner—and you'll breathe less dust, too.

[**TIP**] To help remove sanding dust from your shop, use an ambient air cleaner. They're available in ceiling-mounted or portable models.

Benches and Clamps, page 46

Benches and Clamps, page 46

Gluing and Assembly, page 63

PART TWO

Benches, Clamps, and Assembly

ONE OF THE MOST ESSENTIAL woodworking tools is a workbench. A large, sturdy bench provides a work area for layout, sawing, planing, assembly, and even finishing. The best bench is one you make yourself to suit your size and the type of projects you're most likely to make. Along with the bench you'll need a vise. A large vise will grip the work securely as you saw, pare, and plane. And there are numerous accessories you can make and use in conjunction with a bench for holding your work. Clamps are also useful for securing work to the bench but are mainly needed for assembling parts into finished projects. Assembling projects successfully relies not only on clamps but on choosing the right glue.

Benches and Clamps

FUNDAMENTAL TO ALMOST ALL woodworking is a bench. As you saw, plane, chisel, and scrape, the workpiece will be pushed and pulled and struck. For first-class work, and for your own personal safety, the work must be firmly gripped and supported by a solid bench. This section shows you what to look for in a bench, suggests bench accessories you can make yourself, and also discusses the numerous types of clamps you'll use in conjunction with a bench.

The Workbench

A good workbench should be heavy and of solid construction to resist the pushing and pounding without racking. As you work, your bench continually serves as a reference surface, so it's important that the top be absolutely flat. It should also be thick enough to resist flexing under load. A large, heavy vise is essential, to hold the work firmly for a variety of tasks. Finally, don't forget height. Bending over a low bench is a sure path to a tired and strained back. In contrast, if your bench is too high, you'll be frustrated as you stretch to reach the workpiece, and the quality of your work will suffer. To determine the best height for your bench, stand with your arms at your sides, palms parallel to the floor. The top of the bench should be level with your palms.

If this sounds like a lot of criteria, it's not, really. Through the years, workbench styles have varied as much as the craftsmen who designed and used them. Some are quite elaborate and time-consuming to construct. But a bench does not have to be elaborate to work effectively.

My own bench, for example, is a simple design. Yet it suits me because it's heavy and absolutely resists racking as I saw and plane. And unlike the elaborate yet one-size-fits-all benches available in catalogs, this bench fits my height because I built it myself. When considering a bench, realize that it is most important that your bench fits you and that you have systems in place, simple as they may be, for holding the work.

Steel rods will allow you to tighten your bench as it inevitably loosens.

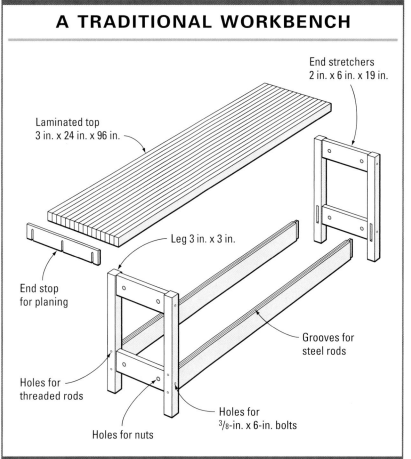

A TRADITIONAL WORKBENCH

End stretchers
2 in. x 6 in. x 19 in.

Laminated top
3 in. x 24 in. x 96 in.

Leg 3 in. x 3 in.

End stop
for planing

Grooves for
steel rods

Holes for
threaded rods

Holes for
$^3/_8$-in. x 6-in. bolts

Holes for nuts

If you prefer to construct a bench of your own, I've included a drawing of mine that you can copy or modify to suit your specific needs. My bench features a 3-in.-thick maple top supported by 3-in. by 3-in. legs. The top is laminated from 6/4 strips on edge to resist warping. The mortise-and-tenon stretcher base is held together with four heavy steel rods that run the entire length of the base. When combined with the wide stretchers, the rods provide a tremendous resistance to racking. As the humidity changes, I can tighten nuts that hold the rods to compensate for shrinkage in the wood.

Bench Accessories

To hold work for sawing, layout, and other tasks, I've equipped the bench with a large iron vise. At the same end of the bench I've added a drop-down stop that holds work for

A large vise is an important part of a workbench.

A drop-down stop is essential for planing.

A drop-down bench stop will accommodate stock of various thicknesses.

A chisel rack keeps tools sharp and close at hand.

planing. The height of the stop is adjustable for thick or thin work. To complete the bench, I built a two-tiered storage rack that attaches to the back edge of the benchtop for holding my large collection of chisels, files, gouges, and squares. The rack keeps these important tools organized and close at hand. The enclosed design of the storage rack also keeps the tool edges sharp and prevents me from getting a nasty cut from inadvertently bumping against a tool.

The classic European bench is a favorite on both sides of the Atlantic. In addition to the standard face vise, European benches also have a tail vise. Unlike that of the classic iron vise, the unique hardware of this vise doesn't obstruct the work. Also, because the jaws open 90 degrees to the face vise, it gives you another position for holding the work.

The movable jaw of the tail vise and the front edge of the bench are fitted with a row of holes for "dogs." Made of steel or wood, bench dogs are useful for gripping stock while allowing full access to the face of the workpiece. Also, unusual shapes that would be difficult to clamp are easy to grip between the faces of bench dogs. Although some woodworkers prefer wooden dogs to avoid damaging edge tools, steel dogs have a firmer, more positive grip. Avoiding the metal near the work is as easy as when using any other metal clamp or vise; simply position the dogs out of the path of the edge tool you'll be using.

Regardless of the style of bench you have in your shop, an assembly bench is a good second bench. The assembly bench should be lower than your workbench to position casework and other assemblies at a comfort-

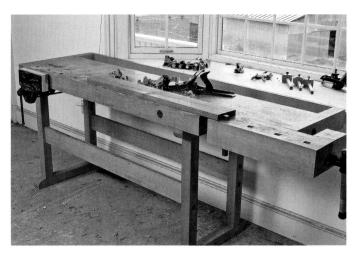

A European bench is outfitted with two vises and a row of dog holes.

A tail vise positions work 90 degrees from a face vise.

Traditional bench dogs are perfect for gripping stock without getting in the way of the work.

This assembly bench is low, for easy access to the work.

able height. The top can be as simple as plywood supported by a framework. If you set the benchtop on folding sawhorses, you can easily knock it down to save space when you're not using it. No workbench design alone can solve all the problems that arise for holding work securely. Although a large

vise is effective for most handwork, you'll need to build a few jigs and accessories as your skills increase.

[TIP] Good lighting is essential to creating good work. I've located my bench in front of a large window for plenty of natural light and a pleasant view.

Drive the tiny escutcheons just below the level of the stock.

Brass escutcheons work well for holding micro-thin stock for planing.

Holding Work for Planing

Probably the most frequent task performed at any bench is planing. As stock is milled to size, shaped, fitted, and assembled, holding it securely for planing becomes more of a challenge.

After bandsawing veneer, you can use small brass nails to secure it for planing. Drive the nails into a scrap of plywood and set them slightly lower than the veneer thickness to avoid striking them with the plane iron. Octagonal stock such as tapered legs and bedposts can be cradled in a V-block. As you plane, use the end stop as you would for any other planing task.

Flush-fitting drawers are typically planed for a precise fit after the dovetails are cut and the drawer is assembled. Attempting to secure the drawer in a vise will cause racking as pressure is applied from the plane; the resulting force can break the joints. Instead, suspend the drawer over a piece of plywood that overhangs the bench. First secure the plywood to the benchtop with clamps and

A V-block will support octagonal work for planing. Cradle the workpiece in the V-block and position it against the bench stop.

To plane the sides of a drawer, suspend it over thick plywood that is clamped to the top of the bench.

A handscrew supports long stock when you're planing.

This bench hook grips long stock for truing an edge.

A wedge locks the board firmly into position.

If you lack a vise, try gripping work with pairs of clamps.

remember to leave the bottom out of the drawer until the fitting is complete.

There are several solutions for gripping long boards for edge jointing. The simplest method is to clamp one end of the work in the vise. The opposite end can be secured in a clamp, and a second clamp will hold the first to the edge of the bench.

Another method is to use a V-block and a wedge. Bandsaw a long taper into a scrap block and save the offcut for the wedge. Clamp the V-block to the benchtop, position the workpiece in the V, and lock it into place with the wedge.

A handscrew clamped to the top of the bench is still one more way to hold long stock on edge for planing. Handscrews are centuries-old clamps with a number of uses. Early handscrews had wooden threads, but today's pivoting steel-thread clamps are stronger and more versatile. And the broad wooden jaws provide a strong, nonmarking grip. Use a second clamp to secure the handscrew to the benchtop.

When milling lumber to size, it's important to realize that lumber is seldom flat, and

► SHOP-MADE SHOOTING BOARDS

These three shooting boards are easy to make from ¾-in.-thick hardwood or plywood scraps.

Shooting Board

Plane against stop

Position hook against edge of bench.

Edge Miter Shooting Board

Plane against stop

Clamp in vise for support.

End Miter Shooting Board

Plane against stop

The bench stop and a pair of blocks clamped to the bench hold this wide board for planing.

so it's necessary to flatten one face before planing it to thickness. As you flatten the stock, it's best to plane at an angle to the board's grain because stock removal goes faster when you're planing slightly across grain. Also, the technique makes it easier to remove distortion across the board's width. To hold the stock securely, position extra stop blocks along one edge to use in conjunction with the end stop on the bench.

The shooting board is another shop-made accessory that has been around for quite some time. It's a simple yet effective jig for holding stock while planing end grain. The jig has two stops; one holds the workpiece and supports the edge to limit chipping of the corner. The second stop holds the jig to the edge of the bench while you're planing. The base of the shooting board supports the edge of the plane. You can also build shooting boards for trimming miters. Although there are specially designed planes for shooting boards, a large bench plane works well, so long as the sides are ground 90 degrees to the sole.

A shooting board holds stock for precise end-grain planing.

Plywood strips screwed to an assembly table hold this door for planing.

After assembling doors, it's usually necessary to plane a few shavings from the rails and stiles to level the joints. Using my broad assembly bench, I fasten strips to the benchtop around the perimeter of the door. The strips secure the door in all directions and yet allow full access to the face of the door for planing.

➤ HOLDFASTS

For several centuries, woodworkers have used forged-iron hooks called holdfasts to secure their work to the benchtop. These simple tools are widely available in woodworking tool catalogs and are still useful today. Holdfasts fit loosely within a hole in the benchtop and hold tight with a wedging action. To use, position the foot over the work and strike the curve of the hook to set it. To release the holdfast, strike the back corner.

Before drilling holes for holdfasts in your prized benchtop, consider carefully the type of work you'll be doing and locate the holes where they will be most effective.

Deep-reach iron clamps can sometimes serve the same purpose as holdfasts. Like holdfasts, they usually work best in pairs. I keep several close at hand when carving or planing moldings.

A traditional holdfast fits within a hole in the benchtop.

Strike the front of the hook to lock it into position.

Striking the back corner will release it.

This deep-reach clamp has plenty of holding power.

To hold wide stock when sawing dovetails, grip one edge in the vise and the other in a hand-screw clamp.

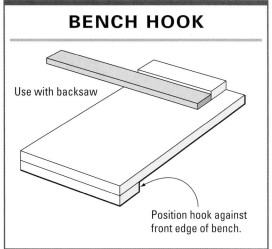

BENCH HOOK

Use with backsaw

Position hook against
front edge of bench.

A bench hook holds stock when you want to saw the shoulder of a tenon.

The bench hook is another classic jig that is similar to the shooting board. It's designed to hold stock for crosscutting with a back-saw. Like the shooting board, it has two stops, one to hold the workpiece and one to hold the jig against the edge of the bench.

Holding Shaped Work

Carvings, turnings, and curved components such as arms and legs for chairs all present unique challenges for clamping. With the exception of the handscrew, most clamps have metal parallel jaws that will easily damage work with curved shapes.

A pipe clamp will hold a leg from the ends while allowing you to shape and refine the curves. Place the clamp in a bench vise to hold the work at a good height and gain the advantage of the weight and mass of the bench.

Typically, if a workbench is at the best height for planing, it is too low for most carving. One easy solution is to place a box in the vise, which lifts the work to a comfortable height for carving. Extend the top

Holding Work for Sawing

When sawing dovetails for casework, you'll want to keep the end of the stock relatively close to the vise to minimize flexing. Secure the other edge of the stock in a handscrew and clamp it to the bench. The work will remain steady with each stroke of the saw, enabling you to precisely follow the layout line.

Pipe clamps are versatile and will easily hold legs for shaping.

This simple box raises carvings and other detailed work to a comfortable height.

To hold odd-shaped work such as this chair arm, cut blocks that conform to the shape of the workpiece.

Some work, such as this finial, may require construction of a fixture.

of the box beyond the sides to provide a position for clamps.

Sometimes the simplest solution for holding odd shapes is to bandsaw a few blocks that conform to the shape of the workpiece. The blocks can then be clamped to the bench without damaging the work.

Finials for casework and other carvings are first turned before carving. To hold the round workpiece for carving I construct a V-block. The work is cradled in the V and gripped between stops. The V-block is mounted in the vise for use while I'm carving.

There are a wide variety of clamps available for tackling every clamping situation.

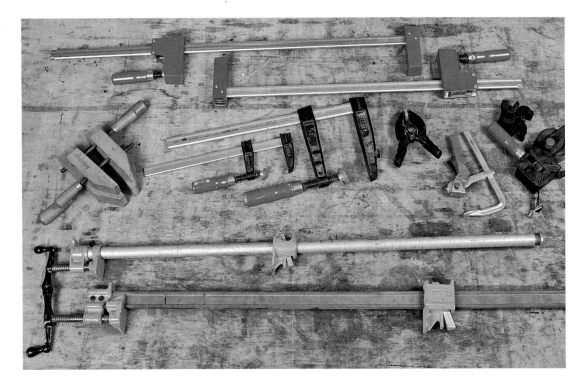

Clamps

The conventional wisdom is that you can never have too many clamps. Not true. Just as for other tools, it's best to make careful purchases, buy just what you need, and use the rest of the woodworking budget for lumber.

Clamps can be used as a third hand for holding stock while you're planing or routing, but they are most often used for gluing up. And although many glues require several hours to fully cure, most only need 20 or 30 minutes of clamping time. After that, you can remove the clamps and use them elsewhere. Also, if you browse the pages of a woodworking catalog, you'll see that there are a lot of different types of clamps on the market these days. However, many of them are specialty clamps you may never need. Here's a rundown of what I find most useful in my shop:

Pipe clamps are an economical favorite for panel, frame, and case assembly.

Pipe Clamps

Pipe clamps are definitely some of the most versatile clamps you can own. You can use them for edge-gluing panels, frame assembly, casework assembly, and even holding table and chair legs for carving. Best of all, pipe clamps are economical compared to their larger, heavier cousins, the bar clamps.

As the name implies, pipe clamps are built upon a length of pipe. The clamp head is purchased separately, and it threads to any length of pipe.

I prefer pipe clamps to bar clamps because they are lighter weight and less awkward to use. Also, the tail stop is easily rotated 90 degrees offset to the head for unusual clamping situations. Bar clamps can't do that. Best of all is the price. You can outfit your shop with at least two or three pipe clamps for the cost of one bar clamp of the same length.

The most common length of pipe clamp needed is 3 ft. Most casework, tabletops, and frames that require assembly will fit within a 36-in. opening. When purchasing pipe, remember to add another 6 in. to the length for the clamp head.

Occasionally, you may need long pipe clamps for large assemblies such as the face frame of a kitchen cabinet. An inexpensive solution is to purchase couplers and attach one or more shorter lengths of pipe end-to-end.

Bar Clamps

Like pipe clamps, bar clamps have many uses. They also have more clamping power than pipe clamps because of the large, I-beam shape of the bar. Yet very seldom do you need all the crushing power of bar clamps. Although I own several bar clamps, I find that I seldom use them because they are heavy and awkward. They're also expensive.

Parallel Bar Clamps

Parallel bar clamps are a European-style bar clamp with several distinct advantages over

You can lengthen a pipe clamp with a coupling and an extra section of pipe.

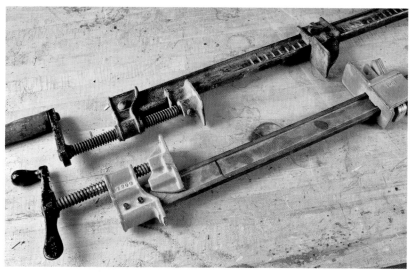

The I-beam bar clamp provides tremendous force for demanding situations.

The square, parallel jaws of K-body clamps distribute pressure evenly.

A T-handle clamp (top) provides a better grip and greater torque than the straight post-type handle (bottom).

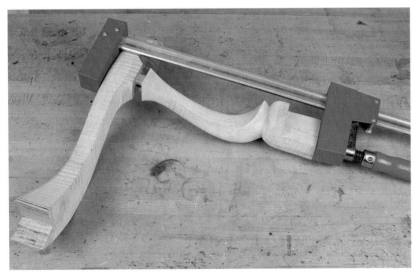

The deep reach of this clamp firmly grips the compound curves of an arm for a chair.

standard bar clamps. The most distinct advantage is that the head and tail stop don't flex under load; they remain parallel so that even pressure is always maintained on the surface of the work. Also, although these clamps are made of steel, the heads are covered with a tough plastic that is gentle to your work. And the deep reach of parallel bar clamps gives them versatility unmatched by conventional bar clamps. Parallel bar clamps are a great choice for doors and other assemblies. However, for edge-gluing boards to create a wide panel, I still prefer pipe clamps; the offset handle makes it much easier to torque them down.

Mini-Bar Clamps

Like their larger counterparts, mini-bar clamps have a stop that adjusts by sliding along the length of a steel bar. These little bar clamps have lots of clamping power, and their scaled-down size makes them extremely versatile. Mini-bar clamps are ideal for

Light-duty bar clamps offer targeted clamping power in tight quarters.

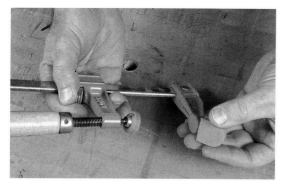

Plastic clamp pads slip onto the jaws to cushion the work and prevent damage.

Ratchet clamps can be quickly repositioned.

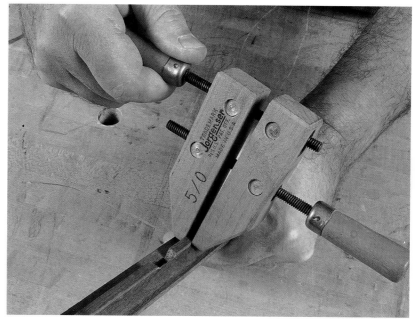

The handscrew clamp has jaws that adjust for various angles.

clamping smaller assemblies, and accessory pads are available to protect the workpiece.

Another version of the mini-bar clamp is the ratchet clamp. The ratchet mechanism of this clamp makes it quick to open and close. Pulling the handle quickly locks the clamp in place; a trigger releases the clamp. This small clamp is ideal for securing work for carving or for any type of work that must be frequently repositioned.

Handscrews

Like the pipe clamp, the wooden handscrew is a perennial favorite. Handscrews have large wooden jaws that distribute the clamp pressure over a broad area. Because of these features, they don't have the tendency to mar the work as some other types of clamps do. The jaws also are adjustable for clamping odd-shaped and angled work. By grasping the two handles and spinning the clamp, you can quickly open or close the jaws. Handscrews are available in a wide range of sizes from the diminutive No. 5/0 to the huge No. 7.

C-Clamps

C-clamps are a good choice for securing jigs, fences, and fixtures to machinery. They have good holding power and plenty of leverage for tightening. The small pads of C-clamps concentrate the pressure in a localized area. This characteristic makes them ideal for

➤ SHOP-MADE CLAMPING JIGS

Sometimes the best clamp for the job at hand is one you make yourself. The clamp in the photo is for securing a turned finial while carving the flame portion. Essentially a V-block, this clamp has stops that are bored to accept round tenons on the turning. One of the stops is adjustable and locks in position with a large wing-screw.

Mitered frames can be awkward to clamp; as pressure is applied, the corners tend to slide out of alignment. A miter clamping jig solves the problem. It pulls the miters together, keeps the joints aligned, and squares the frame, all with one clamp. A row of holes in each of the four beams allows the jig to adjust for almost any size frame. Also, the beams pivot on the clamp blocks to allow clamping of either square or rectangular frames.

Final Clamping Jig

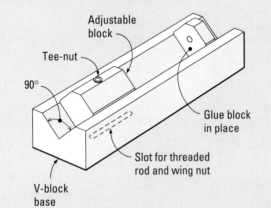

Adjustable block

Tee-nut

90°

Glue block in place

Slot for threaded rod and wing nut

V-block base

Mitered Frame Clamping Jig

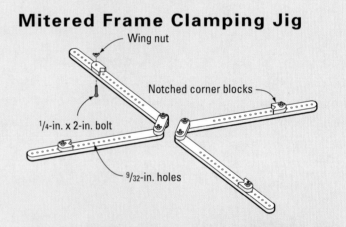

Wing nut

Notched corner blocks

1/4-in. x 2-in. bolt

9/32-in. holes

Threaded rod allows you to make clamp fixtures for unusual shapes.

This shop-made mitered frame jig uses just one clamp to assemble an entire frame.

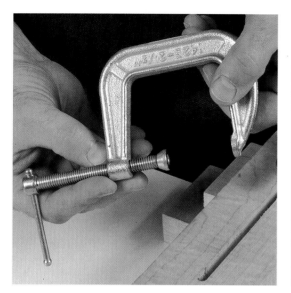

C-clamps are often the best choice for securing jigs to machine tops.

C-clamps have a small footprint that can easily damage the work.

securing jigs to the tops and fences of machinery. However, I avoid using C-clamps for most woodworking applications; the small pads tend to mar the work and the screws are too time-consuming to adjust.

Toggle Clamps

Toggle clamps are designed for securing work in jigs and templates for sawing and shaping. Toggle clamps grip the stock so that you can position your hands a safe distance from the cutting action. The base of the clamp has mounting holes for securing to the jig, and the lever action, rather than a screw, makes the clamp quick to open or close. Toggle clamps are available in a range of sizes and designs, so you're sure to find one for every application.

Toggle clamps can quickly secure stock in jigs and fixtures.

Band Clamps

Band clamps completely encompass the work with a nonmarring nylon strap. After wrapping the strap around the work, use a screw or ratchet to tighten the strap and provide pressure. Band clamps are most useful for clamping unusual shapes, such as octagonal pedestals and other polygons, where it is difficult or impossible to use a standard clamp.

Vacuum Presses

While traditional clamps exert pressure with a screw or lever, the vacuum press uses atmospheric pressure. Although technically speaking the vacuum press isn't a clamp, it is used for clamping. If you enjoy veneering or want to include bent lamination in your woodworking, you'll quickly see the benefits of a vacuum press. With a vacuum press, you won't have to bother with mating forms, clamping cauls, and a ton of clamps. Best of all, the pressure inside the bag is uniform, so you won't be second-guessing about the pressure on various parts of an assembly.

The vacuum press is available in a range of prices to suit most woodworking budgets. Less expensive systems use smaller, lighter-weight bags and continuous running pumps. For more money, you can get larger bags made from more durable materials along with heavy-duty pumps that cycle on and off to reduce wear on the pump.

A vacuum press uses atmospheric pressure to exert force.

After use, the bag rolls up for convenient storage. Once you've tried a vacuum press, you won't go back to the old methods of clamping veneer and bent laminations.

Spring Clamps

Spring clamps work like oversized clothes-pins. A coil spring provides the pressure to small plastic pads. Spring clamps are available in a number of sizes, and they are ideal for small repairs. The plastic pads won't mar the workpiece, and the pads pivot to correspond to the angle of the stock.

The plastic jaws and light pressure of spring clamps make them ideal for small repairs.

Gluing and Assembly

Simple Glue Joints

➤ Rub Joint (p. 70)
➤ Edge Joint (p. 71)

Project Assemblies

➤ Leg-and-Rail Assembly (p. 72)
➤ Mortise-and-Tenon Framework (p. 73)
➤ Mitered Feet (p. 74)
➤ Mitered Box (p. 75)
➤ Dovetailed Casework (p. 76)

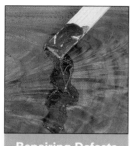

Repairing Defects

➤ Filling a Crack with Epoxy (p. 77)

SSEMBLY IS A CRITICAL time in the process of building a piece of furniture. You may have spent days or even weeks sizing stock, cutting and fitting joints, and shaping parts. Then, in a relatively short period of time, you've got to apply glue, assemble the various parts in the correct order, position clamps, and check for both alignment and squareness. And remember that glue-up is typically irreversible, so you must assemble it correctly the first time. The key to a smooth and stress-free assembly is to conduct a rehearsal of the process, often called a dry run. The dry run is the time to check fit, alignment, and squareness and make any necessary adjustments. It's also the time to determine the number and position of clamps.

Read on to find out how to make your next assembly trouble-free. You'll find solid information on how to assemble your next project and get an idea of the types of clamps you'll need.

No discussion of assembling furniture would be complete without considering the various types of adhesives available today. A trip to the local home improvement center or woodworking store will yield a dizzying array of woodworking glues. While some simply squeeze from the bottle ready for use, others require mixing two parts, and some, such as hot-melt and hide glue, even require heating. As you probably suspect, each type of glue has different working qualities as

Today there are a wide variety of glue choices.

After an hour of drying time, yellow glue easily scrapes off.

expected, the joints to which they are applied should be clean, dry, and well-crafted. More specifically, mating surfaces should touch; adhesives do a poor job of filling gaps. In fact, water-based glues, such as common yellow glue, shrink as the water in the glue evaporates, dissolving any hopes of gap-filling properties. Although some glues, such as polyurethane, don't contract while curing, a thick glueline creates a weak bond. For glues to bond effectively, the surfaces must also be clean, dry, and smooth. The ideal surface is one in which the cells have been cleanly severed by a sharp plane or chisel. A heavily textured, washboard surface created by feeding the stock too quickly through a planer or jointer can limit full contact between the mating surfaces. For the best bond, a sanded surface should be vacuumed or cleaned with compressed air to remove dust particles from the pores of the wood.

Choosing the Right Glue

Glues vary widely in their working characteristics, strength, and temperature range. For example, in cooler temperatures some adhesives slow in their cure rate while others create a weak bond or don't work at all. When using any adhesive I'm unfamiliar with, I always carefully read and follow the manufacturer's instructions. Let's take a closer look at some features to consider the next time you're choosing a glue.

Ease of Use

Glues that can be squeezed from the bottle, spread onto the wood surface, and cleaned up with water top the list as easy to use. White and yellow glues are the most com-

well as characteristics. By having a basic knowledge of the most common types, you'll be able to select the best glue for the job at hand.

Preparing Surfaces for Glue

Despite the technological advancement in adhesives during the past 50 years, there are no miracle glues. For glues to work as monly used woodworking glues not only because they are user-friendly but because they're strong, inexpensive, and emit no harmful fumes. Unfortunately, they also have little resistance to creepage and moisture, and they are not reversible.

Reversibility

A glue's ability to be "unglued" is an important characteristic for guitar makers and furniture restorers. Guitars eventually need repairs because of the tremendous forces exerted by the strings. Repairs to antiques should be made using reversible glue to avoid spoiling the integrity of the piece. However, for most woodworkers, reversibility is unimportant; furniture that is well-crafted and cared for should last at least 100 years or more. The most common reversible glue is hide glue, so named because it is made from ground-up animal skin. However, any glue that is a basic derivative of a natural product can usually be reversed and the parts disassembled. In contrast, modern synthetic glues are not reversible.

Creep Resistance

Any glueline that is subject to excessive stress is subject to creepage. This phenomenon is manifested by a mismatched glueline. Examples include veneer over a solid wood substrate and layers of solid wood glued to create a heavy bedpost. When these broad surfaces move as a result of seasonal changes in humidity, the joints often become slightly misaligned. Glues that have a tendency to creep are also subject to springback if used for bent laminations. An excellent choice for creep resistance is urea resin glue. Polyvinyl acetate glues (PVAs), and white and yellow glues, have moderate creep resistance. Contact cement has very little creep resistance.

Water Resistance

Nowadays, there are several adhesives that offer outstanding resistance to water. Be aware, however, that there are different levels of water resistance among adhesives. It's important to carefully read the label. For example, Type II yellow glue is water resistant but not waterproof, so it's suitable for outdoor furniture but not for boats. Resorcinol and epoxy are both waterproof and the best choices for work that will be subjected to long periods of submersion. These adhesives come as two parts that must be measured and mixed prior to application.

Toxicity

In general, two-part glues that require mixing before use are the most toxic. These glues cure by chemical reaction, and the chemicals release harmful fumes. It pays to

Before you begin, gather the tools you'll need.

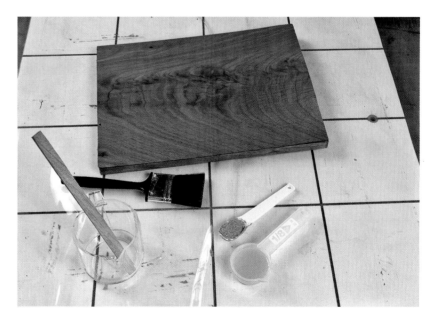

The best glue applicator is one that enables you to spread the right amount in the right location.

You'll gain better control with this flux brush if you first trim the bristles.

Mix two-part glues thoroughly.

be safe. Read the manufacturer's label and take the necessary precautions to protect yourself.

Tools for Spreading Glue

When spreading glue, it certainly helps to have the right tool for the job; you'll get even coats of the right amount of glue in the right location. A paint roller works well for large jobs, such as when you laminate planks to make a benchtop or spread glue on a substrate when veneering. A short roller is usually large enough and will give you good control of the spread. For smaller applications, an old credit card or phone card works well.

A very simple yet effective tool for glue application is a thin, tapered stick. The stick will easily reach into mortises, and you can use it as a small paddle for coating the mortise walls. A flux brush also works well for reaching into tight spots. However, the bristles are too long for good control, so I trim the ends before putting it to use.

Inevitably you'll need to make a small repair, and a syringe will squeeze just the right amount of glue into any crack or crevice. The ones I use are available at farm supply stores. The large needle will allow glues to flow through.

Glue-Up Strategies

Gluing up is a high-risk venture, and there's definitely a lot at stake. Perhaps you've spent days or even weeks selecting, preparing, and joining stock. Now it's time for assembly. Because glue is typically irreversible, you've got to get it right the first time. You've got only a few minutes to work before the glue sets up, and yet everything must be aligned and checked for square. It sounds stressful, but it doesn't have to be. Glue-ups can be calm, and everything can go as planned. The key is to have a plan. I seldom glue anything without first performing a dry run. Think of a dry run as a dress rehearsal. It's a time to practice the movements as you assemble the various parts and position the clamps. A dry run will also allow you to check all joints for fit and check the entire assembly for square. If there are any problems, they will be apparent when it's easy to fix them. If there are no problems during the dry run, then you're ready to apply the glue. All the tools are in place, and the clamps are adjusted for the size of the work. Here are some other criteria that are important to a dry run, as well as to the glue-up itself.

- **Work with square stock:** Stock for glue-up should be flat, true, and square. For example, if you're gluing boards edge-to-edge to make a tabletop, the face of each board should be free of warp and the edges should be true (straight) and 90 degrees to the face. You can square

A low-angle block plane will cleanly level the end-grain pins of a dovetail joint.

the stock with a jointer and planer or with hand tools.

- **Make joints that are tight:** Complex joints, such as the dovetail and mortise-and-tenon, should fit with hand pressure or gentle coaxing with a mallet. Remember, the purpose of clamps is to keep a joint closed until the glue sets, not to close poorly fitted joints. Joints that require excessive clamp pressure to close will continually be under stress once the clamps are removed. Under these circumstances, the joints will ultimately fail.
- **Use minimal clamp pressure:** Assuming that the joints are well-crafted, remember that most clamps can apply much more pressure than is needed to keep a joint closed. Too much pressure can crush the wood fibers and rack an assembly out of square. As you're tightening clamps, use only enough pressure to close the joint.
- **Use sub-assemblies:** It's much easier to glue up just a few parts at a time. Attempting to glue together an entire project, even a small one, can be an

Even a small project should be glued up in sub-assemblies.

During the Glue-Up

Before you disassemble the dry run, note the number and position of clamps. Gather tools that you'll need for the glue-up, such as squares, pinch rods, glue applicators, and a dead-blow mallet to gently coax parts into position.

[TIP] When clamping boards edge-to-edge, I alternate the clamps over and under to equalize the pressure on both faces of the stock. Otherwise, you can inadvertently cup the panel during glue-up.

---

invitation for failure. It's difficult to spread glue on all the parts before the glue begins to "skin over." Also, the project may buckle and distort under the weight and pressure of so many clamps. Instead, glue together sub-assemblies, such as the left and right sections of a small table. Then, once the glue has dried, glue together the entire table. The dry run will make you aware of whether you're attempting to assemble too much at once.

- **Glue up on a flat surface:** Even though your joinery is flawless, it's still possible to glue a twist into a paneled door or dovetailed drawer if the work surface is distorted. I prefer to glue assemblies on the top of my bench because I know that it is flat. I'll sometimes use sawhorses when gluing large work, but I'll first make certain that the top of each sawhorse is positioned in the same plane. If necessary, I'll shim one leg of a sawhorse.

Spread the glue evenly on all long-grain surfaces of a joint and methodically assemble the parts in the same order that you used during the dry run. Once clamp pressure is applied, I like to see just a bit of glue squeeze out. This is good insurance to let you know that the joint isn't glue-starved. I like to avoid excess squeeze-out, though; it makes a sticky mess and is time-consuming to remove from the surfaces, corners, and crannies of the work. It takes practice to know how much glue to apply, however. When you're first starting out, it's best to err on the side of too much.

The best time to remove glue is after it has partially set up. At this stage, the glue is no longer liquid but has a soft, plastic texture and will easily scrape away. Wiping the wet glue with a damp rag will push it into the surrounding pores and haunt you during finishing. Allowing excess glue to harden is problematic, too. Hard, dried glue is tenacious; as you scrape it away, it will often chip out small areas of the wood surface. Ouch!

▶ DIRECTING THE CLAMP PRESSURE

Sometimes assemblies that seem straightforward end up being difficult once you get the glue and clamps on. That's why a dry run with clamps is important. Study the joint to determine where the pressure is needed and use clamp blocks to direct the pressure. For example, on a typical leg-and-rail assembly, the pressure is needed behind the rail, not across the entire leg. Pressure applied to the entire leg will twist it out of square, yet a simple clamp block will distribute the pressure correctly. Some joints, especially on corners that are not square, require a more customized clamp block. Dovetail joints require notched clamp blocks that direct clamp pressure to the tails.

Clamp Block for Dovetails

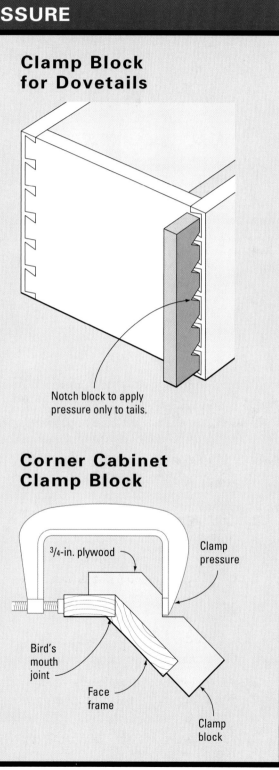

Notch block to apply pressure only to tails.

Clamping a Leg-and-Rail Assembly

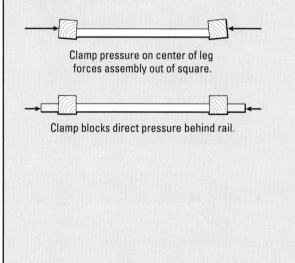

Clamp pressure on center of leg forces assembly out of square.

Clamp blocks direct pressure behind rail.

Corner Cabinet Clamp Block

3/4-in. plywood

Clamp pressure

Bird's mouth joint

Face frame

Clamp block

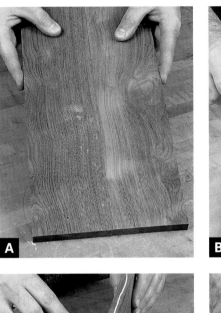

A

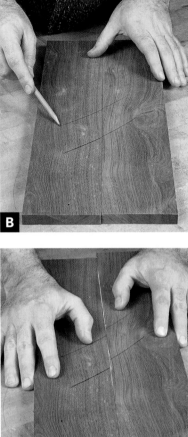

B

Rub Joint

A rub joint is simply an edge-to-edge joint that uses no clamp pressure. This technique works well for small joints and/or awkward assemblies in which it's difficult to position a clamp. The edges must fit perfectly with no gaps or spring in the joint **(A)**. Mark the position of the stock with a pencil to make realignment easier **(B)**. Next, spread the glue evenly along one edge, **(C)** and rub the joint back and forth a couple of times to even out the film of glue **(D)**. Align the joint on the marks you made previously, and allow the glue to dry **(E)**.

C

D

E

Edge Joint

For several years I lived near a professional logger and would often purchase large hardwood logs from him. Another neighbor with a portable bandsaw mill would saw the logs to my specifications. Consequently, I have a large stockpile of walnut, cherry, maple, and poplar boards, many as wide as 2 ft. These beautifully figured boards are ideal for small tabletops, door panels, desk lids, or anywhere I want a show of eye-catching wood. I don't like to use more than one board for these surfaces; the seam is just too distracting. Even so, there are times when I must glue together two or three boards to make a large tabletop or bottoms for drawers.

When gluing boards edge-to-edge, I don't use splines, dowels, biscuits, or any other device to add strength or help with alignment. An edge-to-edge joint comprises long grain, and so it's already stronger than the surrounding wood. And because the boards were flattened on the jointer before thickness planing, alignment is a cinch. As I apply clamp pressure, I just pull or push the joint into alignment.

Begin with a dry run **(A)**. Check the boards with a straightedge to be certain that the assembly is flat **(B)**. Next, spread an even coat of glue along one edge only **(C)**. Too much glue lubricates the joint, causing the boards to slide out of alignment as clamp pressure is applied. Start clamping in the middle of the panel and work toward each end. Before you tighten the clamp, check the alignment in that area with the tip of your finger **(D)**. If necessary, push the boards at the ends to align the joint. As you continue clamping, position the clamps over and under the assembly to equalize clamp pressure **(E)**. Finally, snug each clamp slightly and set the work aside to dry **(F)**.

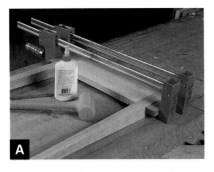
A

B

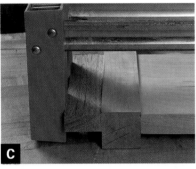
C

D

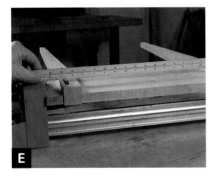
E

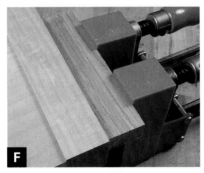
F

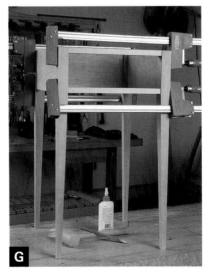
G

Leg-and-Rail Assembly

Leg-and-rail construction is commonly used on tables, chairs, beds, and even some forms of casework. The legs are often two or three times as thick as the rails, which produces a challenge during clamping. If clamp pressure is applied across the entire surface of the leg, it racks the leg out of square in relation to the rail. (See the sidebar on page 69.) The solution is to use clamping blocks. Because the blocks are the same thickness as the rail, they focus the clamp pressure only to the joint area, which prevents racking. Even so, it's best to check the legs to be certain that they're in the same plane.

Begin by gathering your tools and performing a dry run **(A)**. Check the joints for closure, and, if necessary, slightly undercut the tenon shoulder to close the joint. Next, coat the mortise walls and tenon faces with glue **(B)**. Position the clamp blocks on the legs directly behind the joint **(C)** and apply moderate pressure to close the joint **(D)**. Now, invert the assembly and position a straightedge across the inside surfaces of the legs **(E)**. These surfaces should lie in the same plane. If necessary, reposition the clamp blocks slightly to correct the problem. Allow any glue that squeezes from the joints to dry until it is firm but not hard **(F)**. Then carefully pare it away with a chisel. After the left and right assemblies have dried, glue together the entire table **(G)**.

Mortise-and-Tenon Framework

Mortise-and-tenon framework has broad applications in woodworking. The framework is commonly used as a face frame on casework. By adding a panel to the framework, you've created a door. The long frame members are called stiles and are mortised to accept the tenons on the ends of the horizontal members called rails. The framework in this photo-essay will be used as a base under a chest to attach the bracket feet.

Notice in the photos the extra length of the stiles. This is a solid, tried-and-true method for building frames. The extra length, called "ears," is left intact until after assembly, at which time it is sawed off. The ears serve several purposes. They add strength to the stile when you're mortising, and they eliminate the need for absolutely perfect alignment during glue-up. Instead, the ears are sawn flush with the rails. Also, when you're disassembling a paneled door after the dry fit, the ears provide a spot to tap open the joints without damaging the door.

After a dry run, coat the walls of each mortise **(A)** and the face of each tenon **(B)** with a film of glue. Assemble the frame and apply a small amount of clamp pressure to hold the joints firmly in place **(C)**. Finally, check the frame for square **(D)**. I'm using an accurate framing square (not all framing squares are 90 degrees), but pinch rods work well, too. If necessary, loosen the clamps slightly, tap the end of a stile with a dead-blow mallet to square the frame, then retighten the clamps **(E)**.

Mitered Feet

Ogee feet must support the weight of the case-work on which they are attached. So for additional strength and help with alignment, I use a ¼-in. plywood spline. Before glue-up, I bandsaw the bracket profile; after glue-up, I use the bandsaw to saw the ogee contour in the faces of the foot. Remember to offset the spline toward the inside so that you can avoid inadvertently cutting into it and spoiling the foot.

Small bar clamps work well for gluing up feet. These mini-clamps are available in different lengths, provide plenty of clamping pressure, and are not nearly as awkward as larger clamps.

I start with a dry run to adjust the clamps and do a final check on the fit of the joint **(A)**. Next, I apply a thin, even film of glue on both halves of the joint, including the spline grooves **(B)**. I begin assembly with the spline and gently push the joint together. As I position each clamp, I start with a small amount of pressure **(C)**. Too much clamp pressure on the first clamp will misalign the joint, in spite of the spline.

Once all the clamps are in place, I tighten them each once more. If the miter is slightly mis-aligned, loosen one clamp and gently tighten the opposing clamp to bring the joint back into alignment **(D)**.

Mitered Box

Mitered boxes can be tricky to clamp; the joints don't interlock, and they easily slip out of alignment as clamp pressure is applied. Small boxes are especially difficult because there is little area in which to position clamps. One sure method is to assemble the box with packing tape (**A**). The sticky tape holds surprisingly well, and it has enough tensile strength to apply sufficient pressure as the glue dries. The tape is available in different colors, but I prefer the clear as it allows me to view the assembled joints. This is a neat trick, it works well, and it's fun. As with most gluing operations, try this technique first without glue to get a feel for it.

Begin by unrolling a length of tape on the bench, sticky face up (**B**). Apply glue to the miters (**C**) and stick them to the tape end-to-end (**D**). Now simply fold up the box (**E**). When you get to the last corner, stretch the tape to apply pressure to the joints and stick it to the box (**F**). Now set the box aside and allow the glue to dry (**G**).

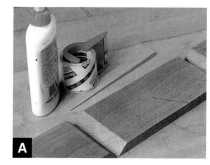

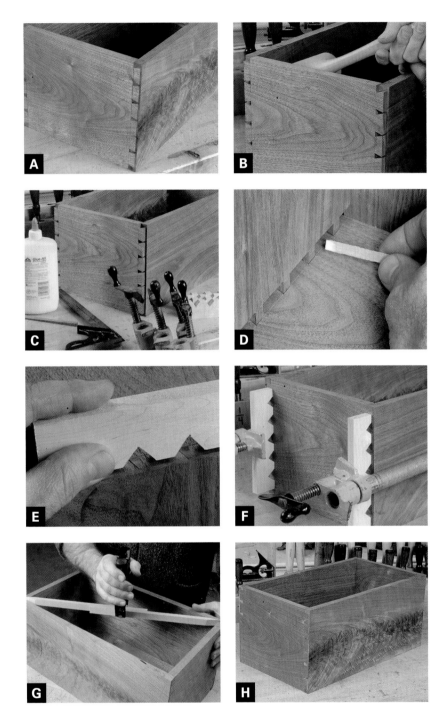

Dovetailed Casework

Dovetails, especially those that are cut by hand, typically only fit one way. Gluing together a dove-tailed box can be like solving a puzzle as you search for mating parts. To avoid trial-and-error fitting or frantically searching for labels as the glue sets up, I use a different approach. I begin by assembling the box and checking for fit and squareness **(A)**. Next, I use a dead-blow mallet to gently tap the joints partially open but not completely apart **(B)**. This allows me access to the inner workings of the joint to apply glue, yet I can easily tap the joints closed again. Next, I gather the tools I'll need for glue-up, including clamps, clamp blocks, and a square **(C)**. Pinch rods work well, too, for checking boxes for square, especially large casework. To spread the glue without making a sticky mess of the box, I use a thin, tapered stick. Quickly and carefully I apply glue to the long-grain surfaces of each pin. **(D)**. After driving the joints home with careful, gentle taps of the mallet, I position the notched clamp blocks over the tails **(E)**. A moderate amount of clamp pressure will keep the joint closed tightly until the glue sets **(F)**. Remember to check for square **(G)**. If necessary, gently push on the acute corners to square the box. Finally, set the box aside on a flat surface until the glue has completely dried **(H)**.

Filling a Crack with Epoxy

The most beautiful, highly figured wood often has the greatest number of natural defects, such as knots and stress cracks **(A)**. An easy way to repair loose knots and other defects is with epoxy. Mixing sanding dust with the epoxy will easily conceal the repair. Squeeze equal amounts of clear, five-minute epoxy onto the mixing board **(B)**. Mix the two components **(C)**, then mix in a small amount of sanding dust **(D)**. Apply the epoxy with a stick and completely fill the void. After the epoxy has fully cured, scrape away the excess and sand it level with the surrounding surface.

> ⚠ **WARNING** Two-part epoxies give off fumes that may be harmful, so use a respirator or provide adequate ventilation.

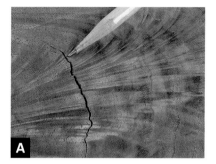

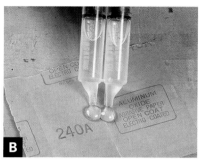

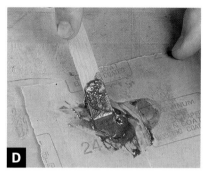

PART THREE

Hand Tools

PERUSE THE PAGES of almost any woodworking catalog and you'll see a dizzying array of hand tools. At the same time, there are more affordable choices in woodworking machinery than ever before. So why use hand tools? Part of the appeal is certainly the romance. Hand tools are a pleasure to use, and they leave behind a distinct surface texture that says "handmade." Using hand tools is skilled, quiet work that provides a great sense of personal satisfaction and accomplishment. The fact is, there are many times when a hand tool is also the most efficient choice. Setting up a machine for a complex cut can gobble up time. Additionally, you can create fine detailed work with hand tools that is often difficult or impossible to produce with a machine.

Measuring and Marking Tools

"MEASURE TWICE, CUT ONCE" is still good advice; nothing is quite as disappointing as miscutting a prized board. But accurate layout is just as important. Layout provides a critical road map to give you direction and sequence for each series of cuts. Layout is the process of measuring and marking the location for joints, curves, and other important details.

Measuring Tools

In any shop, measuring tools should be close at hand. My favorite measuring tool is still the classic Lufkin x46® folding wood rule. It will measure up to 6 ft. (quite a bit larger than many woodworking projects), yet it quickly folds up and fits in a hip pocket. Unlike a steel tape, a folding rule lies flat on the work and doesn't inadvertently retract during use. It also doubles as a depth gauge for cutting mortises. Put a drop of oil on

each joint every few months, and this perennial favorite will provide years of service.

When rough-cutting lengths of lumber, I reach for a steel tape measure. A 20-ft. tape provides plenty of length, and the steel is stiff enough to extend it several feet without flexing.

For dividing or transferring measurements (without math!), dividers are often the most accurate and efficient choice. Their steel legs pivot under spring tension and are held in position with an adjustment screw. Dividers are especially effective for layout of curved and irregular objects such as carvings. Where a rule or tape just isn't practical, dividers are available in several sizes ranging from 4 in. to 12 in. As the points get dull, sharpen them using light strokes with a mill file.

Spring calipers are used for measuring diameters when you're turning, carving, and sculpting. As with dividers, a simple adjust-

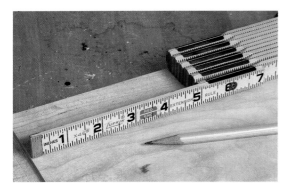

The 6-ft. folding rule is versatile and compact.

The steel tape is a great choice for measuring long lengths of lumber.

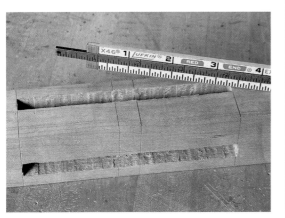

The slide on a folding rule is ideal for measuring the depth of a mortise.

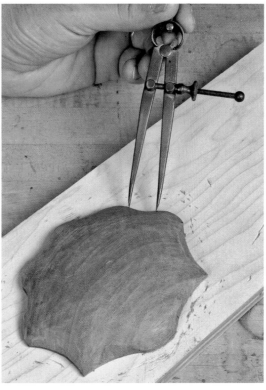

A divider is the best tool for layout of carvings.

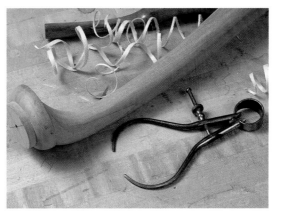

Use the spring caliper to gauge carvings and turnings.

ment screw keeps the spring under tension. When turning, I may use a half-dozen different calipers so that I don't lose efficiency by continually resetting the same pair.

When fitting joints or machining parts to precise dimensions, you'll find a machinist caliper useful. They're available in both analog and digital readout. Equipped for both inside and outside measurements, a machinist caliper will measure in hundreds or even thousandths of an inch. Although this degree of precision isn't needed for most woodworking, it's useful for milling wood to exact thickness or setting up a table saw to cut tenons for a precise fit within a mortise.

A dial caliper is useful for making precise measurements.

An electric stud finder is invaluable when you're hanging shelves and cabinets.

Once you have a machinist caliper in your shop, I'm sure you'll find other uses for it.

For hanging cabinets or shelves, a stud finder will help you ensure that they stay put. This small, handheld device uses electronics to locate the edges of a wall stud. It's also much kinder to your walls than the old method of driving nails until you hit something solid. Simply turn on the device and slide it along the wall until it lights up and emits a tone.

Squares and Bevel Gauges

The square is one of the most essential woodworking tools. My favorite is the machinist's combination square; it's highly accurate and extremely versatile. A sliding iron head locks in place anywhere along the length of the steel rule. Both the head and the rule are carefully machined for accuracy. The numerical markings and divisions are engraved in the face of the rule to eliminate wear, and the best rules have a satin finish, which is much easier to read. It's the sliding head that makes a combination square so

A combination square also works as a gauge for marking parallel lines.

Rules made of satin chrome (bottom) are easier to read than ones made of polished steel (top).

The sliding head of the combination square can be positioned for inside corners.

versatile. When you move the head to a new position, the square functions as both an inside and outside square. However, it's also useful as a depth gauge for checking mortises and other joinery, a height gauge for machine setup, and a 45-degree square for checking miters. The large, 12-in. combination square is the most flexible, but it's helpful to have several smaller sizes. The 6-in. square will fit into smaller spaces, and the 4-in. easily slips into your pocket for use all around the shop, including layout and checking joinery.

When using a combination square, always remember to lock the head to the rule after making adjustments. And avoid the inexpensive hardware-store squares. Often constructed of aluminum, plastic, and stamped steel, these poorly made varieties are no bargain. In fact, they are often not even square. The drawing shows how to check the accuracy of any square.

To check the depth of a mortise, use a combination square.

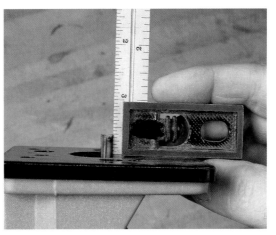

This small square is ideal for setting the depth of a router bit.

A combination square is also effective for checking miters.

CHECKING A SQUARE

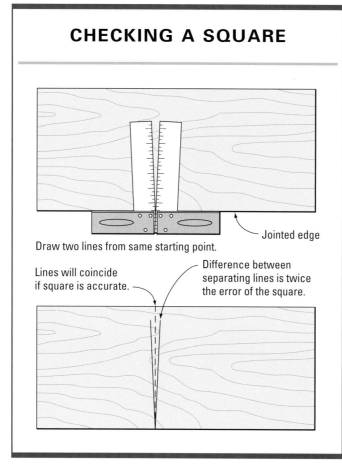

Draw two lines from same starting point.

Jointed edge

Lines will coincide if square is accurate.

Difference between separating lines is twice the error of the square.

You can also purchase accessories to give your combination square even greater versatility. The center head makes it quick and easy to locate the center of both round and square stock. And the protractor head rotates 360 degrees to easily locate angles other than 90 degrees.

Although intended for rough carpentry, the framing square is useful for large layout work as well as for checking large assemblies for square. As with all layout tools, be sure to purchase a quality framing square. Although you can make minor adjustments to a framing square that isn't quite 90 degrees,

Identify specific angles easily with a protractor head.

A center head will accurately locate the center of round or square stock.

ADJUSTING A SQUARE

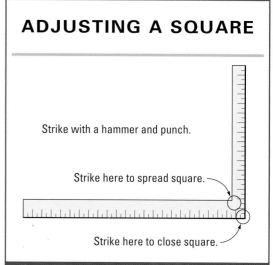

Strike with a hammer and punch.

Strike here to spread square.

Strike here to close square.

a cheap square isn't worth the trouble. Finally, for accurate machine setup, a small machinist's square is handy. These little squares are precisely machined, and the blade is securely pinned to the head.

For laying out angles other than 90 degrees, the sliding bevel is the tool of choice. Like a square, the bevel has a blade and head, but the joint that fastens the two pivots and locks them in place at any angle. You can set the bevel using a protractor or by graphing out a ratio of whole numbers, such as one to six, as shown in the drawing.

The dovetail joint has been a mark of craftsmanship for centuries. Today, despite myriad router jigs, there are still good reasons

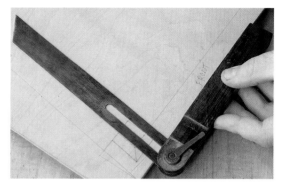

A bevel gauge will adjust to transfer any angle.

The bevel gauge is available in a number of sizes.

A small machinist's square is handy for squaring blades and bits.

SETTING A BEVEL GAUGE

Ratio of Whole Numbers

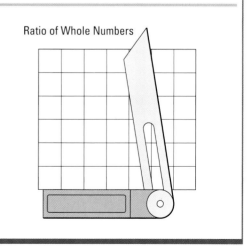

This inexpensive gauge is the perfect tool for dovetail layout.

For the most precise marks, sharpen a pencil to a chisel edge.

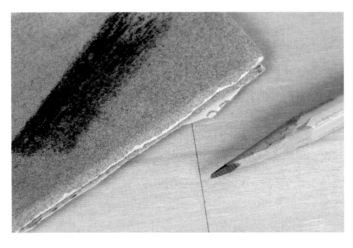

Thin plywood patterns make it easy to duplicate curves.

for cutting dovetails by hand—for nothing but the pleasure. Although you can lay out dovetails with a sliding bevel, the dovetail gauge is preset for marking the angles of the dovetails. Designs vary widely, but my favorite is a simple, inexpensive aluminum extrusion. It's available in two different angles, and the soft aluminum allows you to easily alter this basic tool to any angle you may need.

Marking Tools

Whether you're rough-cutting stock to length or laying out fine joinery, most layout is done with a pencil. Pencils are inexpensive, readily available, and obviously, you can erase a pencil line if needed. To increase the accuracy of a pencil, try a hard No. 4 type rather than the standard No. 2, which is quite soft. For even greater precision try sharpening the point to a chisel edge with a scrap of sandpaper. This old draftsman's trick results in a finer, more precise layout line. The standard No. 2 pencil has a soft lead that is ideal for laying out curves for bandsawing. For layout on dark woods, you can find white pencils at art supply stores.

For the most precise layout, use a knife.

It's easy to set this marking gauge because it's graduated.

Their soft, waxy core isn't sharp enough for precise joinery layout, but they're great for outlining curves for bandsawing or scrollsawing.

For the greatest accuracy during layout, I use a knife. Woodworking catalogs offer elaborate knives with polished blades and exotic wood handles, but I prefer an inexpensive modeling knife. The blade is thin for reaching into tight areas, sharp, and very precise. As the blade dulls I simply toss it out and install a fresh one.

Marking Gauges

The marking gauge is a centuries-old tool that is still one of the most useful. It consists of a wooden head that slides along the length of a wooden beam and locks in place with a thumbscrew. A steel marker fixed to one end of the beam scribes a permanent line with or across the grain. The marker can be a round pin, a small knife, or a tiny wheel. My favorite marking gauge is the Stanley No. 65®. Long out of production, it can still be found where old tools are sold. It features a graduated boxwood beam with a

The pin of this gauge can be removed for sharpening.

hardened steel pin that easily removes for sharpening. All the wear points are fitted with brass.

When correctly sharpened, a gauge should scribe a crisp, sharp incision with or across the grain. The conical point of pin gauges tends to tear the wood fibers instead of scoring them. You can correct this by grinding a chisel profile on the pin and honing it as you would any chisel. I avoid wheel gauges because most create a soft indentation rather than a crisp incision.

When shopping for a marking gauge, you may want to look at the old ones. Like most old tools, old marking gauges have useful features not found on many new tools sold today. For example, most older gauges have a graduated beam. The graduations make setups much faster and far more accurate. Some old gauges have an opening at the opposite end of the beam for use with a pencil. Others feature a special face for scribing curved work.

Tools for Circles, Curves, and Arcs

Remember that cheap, pressed-steel compass from fifth-grade math class? It's time to replace it with something sturdier. A good compass has forged steel removable legs and a thumbwheel for fine adjustments. If you replace the pencil with a steel leg, this versatile tool doubles as a large divider.

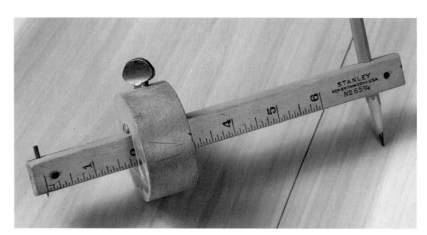

This gauge holds a pencil at one end and a tiny knife at the opposite end.

This special gauge is designed for marking a line parallel to a curve.

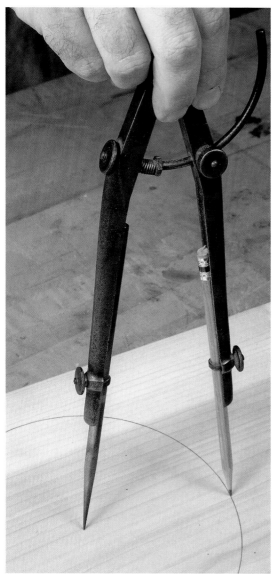

A compass is used to mark circles and arcs.

Trammel points clamp to a stick used to draw large arcs and circles.

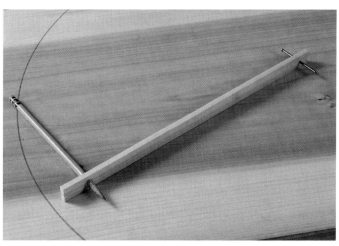

A stick fitted with a nail and pencil makes an economical trammel.

To lay out large circles and ellipses for tabletops, I reach for a pair of trammel points. This classic tool clamps to a stick to lay out circles beyond the reach of a compass. If you need just an occasional circle, a homemade trammel may suffice. Just drill a hole at each end of a stick, one sized to accept a nail and the other a pencil.

Drawing free-form curves is best done with the help of a flexible stick. Simply flex the stick along several predetermined points and trace along it with a pencil.

An awl is a good tool for marking the center of a circle before drilling or locating centers of lathe work. Its sharp point creates a positive indentation that prevents the drill from skipping off-center. Awls are very simple tools, so an inexpensive one will suffice. As the point dulls, sharpen it with a few strokes of a mill file.

Don't be tempted to use your awl as a scribe for scoring layout lines. The conical point will tear the fibers and make it difficult to achieve tight-fitting joinery.

To mark free-form curves, try flexing a thin stick between brads.

An awl will accurately locate the center point for drilling.

▶ SMART LAYOUT PROCEDURES

Careful measuring and marking, or layout, is one of the most essential aspects of woodworking. When I've cut a joint in the wrong location or built a door that doesn't fit the opening, I can usually trace the error to the layout. I like to think of layout as a road map; it points me in the right (or wrong) direction. Here are the steps I use to get it right:

- *Measure twice, cut once.* This old saying is still good advice. Always double-check your measurements.

- *Measure the longest length first.* For example, if you're making a chest of drawers, measure the chest height first, then mark the smaller measurements, such as the drawer locations, moldings, and feet.

- *Transfer the layout.* After the initial layout is made, transfer it to the mating pieces. This ensures that all pieces correspond. For example, when making a table, lay out the mortise location on the first leg, then transfer the measurements around the corner to the adjacent surface. Finally, clamp mating legs to the first and transfer the layout. This method avoids the inevitable errors that can occur when you measure each layout separately.

- *Use a knife for handwork.* You can follow a knife line with a handsaw or chisel much more easily than you can follow a pencil line. Pencil lines have thickness, which leads to inaccuracy. Yet a chisel will easily follow an incision made from a knife.

- *Keep the layout tools sharp.* Whether using a pencil (appropriate for some types of layout) or a knife, don't allow it to become dull. Stop during layout and resharpen as the pencil lines widen or the knife begins to tear the wood.

- *Watch the reference surfaces.* Layout tools have reference surfaces, such as the head of a square or the face of a marking gauge. For accurate layout, the reference surface must stay in contact with the workpiece as you strike your marks or lines.

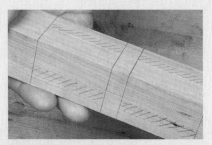

Strike layout marks on one face first, then transfer the marks to adjacent faces.

Clamp like parts together and check that the ends are even.

Transfer marks from one piece to another like piece.

Maintain constant pressure as you strike lines with a marking gauge.

Drafting triangles are inexpensive and accurate.

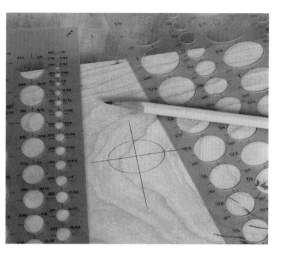

Drafting templates are available for drawing common geometric shapes.

A French curve will enable you to readily create free-form curves.

This simple shop-made tool will easily trace compound curves.

Drafting Supplies

Every good design begins with a drawing; a trip to the office supply store will yield a number of inexpensive plastic drafting tools useful for woodworking layout and design. Triangles are available in 45-degree/ 45-degree/90-degree or 30-degree/ 60-degree/90-degree and in a number of sizes. There are also a number of templates available for drawing small circles and ellipses. After you've sketched your design, you can smooth it out with a French curve. These curvaceous templates are available in several sizes, and they make drawing beautiful curves within everyone's reach.

You can easily copy existing curves from an antique with a tracing tool. This useful little tool is a snap to make. The blunt point follows the curves of a three-dimensional object while the pencil duplicates the existing outline.

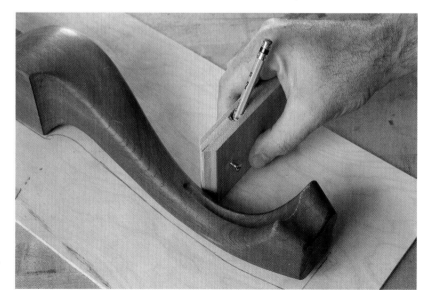

► GEOMETRIC CONSTRUCTIONS

Much of furniture design and layout is drawing geometric shapes. Of course, layout of squares, rectangles, and circles is relatively straightforward. Many shapes are interrelated; for example, an octagon is drawn from a square, while an ellipse is drawn from a rectangle. The following drawings show how to lay out some of the most common shapes you'll use in building a lifetime of furniture.

Drawing an Ellipse

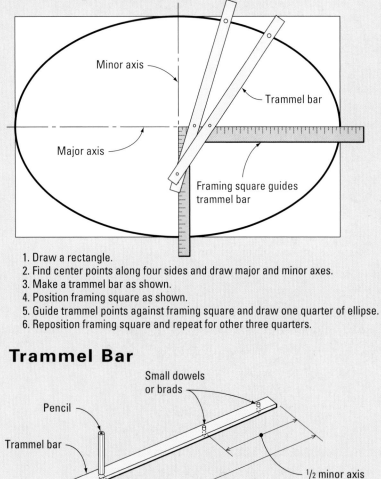

Minor axis

Trammel bar

Major axis

Framing square guides trammel bar

1. Draw a rectangle.
2. Find center points along four sides and draw major and minor axes.
3. Make a trammel bar as shown.
4. Position framing square as shown.
5. Guide trammel points against framing square and draw one quarter of ellipse.
6. Reposition framing square and repeat for other three quarters.

Trammel Bar

Small dowels or brads

Pencil

Trammel bar

¹⁄₂ minor axis

¹⁄₂ major axis

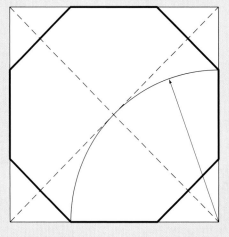

Drawing an Octagon

1. Draw a square.
2. Draw diagonals to locate center point.
3. With a compass, draw arcs from corners through center point.
4. Connect points at corners.

Trammel points are adjustable so you can draw figures in a variety of sizes.

Drawing Ogee Curves

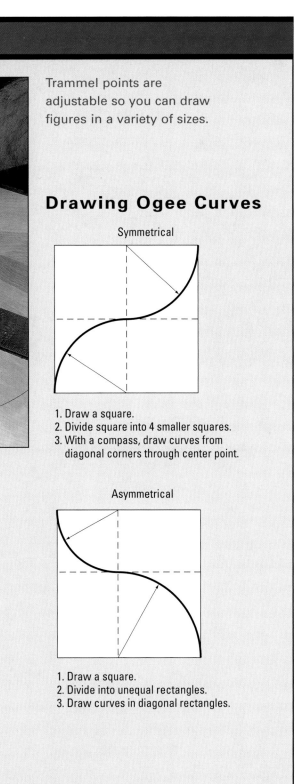

Symmetrical

1. Draw a square.
2. Divide square into 4 smaller squares.
3. With a compass, draw curves from diagonal corners through center point.

Asymmetrical

1. Draw a square.
2. Divide into unequal rectangles.
3. Draw curves in diagonal rectangles.

CONTOUR TRACER

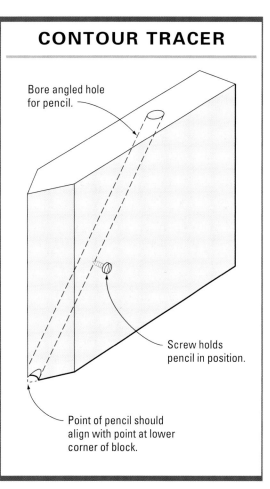

Bore angled hole for pencil.

Screw holds pencil in position.

Point of pencil should align with point at lower corner of block.

After you have painstakingly created your design, you can store the data for the next layout with a story stick and patterns. A story stick is a full-size record of all the horizontal or vertical measurements in a piece. Once drawn, the story stick eliminates math and layout errors. It's especially useful for making multiple pieces and works effectively for everything from small turnings to large casework. Thin plywood patterns make it easy to duplicate curves, and the edges won't become ragged or torn like those of a paper pattern. I like to write dimensions, construction notes, and other data right on the patterns so I'll have this important information close at hand and permanently recorded.

A story stick contains linear measurements for each layout.

Patterns store valuable design information for easily duplicating parts.

It's always useful to have a few precision straightedges available for checking surfaces of tools and machines. Look for those with precisely machined edges; cheap examples will have rough, stamped edges.

Whenever you're flattening wide boards with a plane, it's helpful to have a pair of winding sticks. Positioning a contrasting stick across each end of the board makes it much easier to spot twist in the board.

While you're making winding sticks, mill a couple of extras and point the ends, for use as pinch rods. This simple tool works from the principle that diagonal measurements are equal when the corners of any rectangular box are square. Hold the pointed rods in opposite corners; pinch them together at the overlap, and then check the adjacent corners without releasing your pinch. (You can also use a spring clamp to hold them together.) The difference, if any, is twice the error.

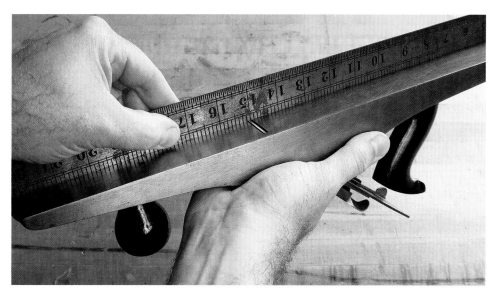

A reliable straight-edge is widely useful for layout and checking accuracy.

Make winding sticks from contrasting woods.

MEASURING DIAGONALS WITH PINCH RODS

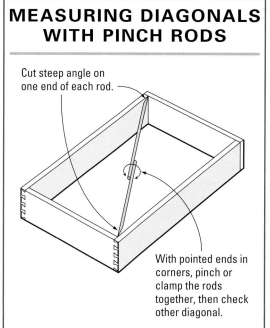

Cut steep angle on one end of each rod.

With pointed ends in corners, pinch or clamp the rods together, then check other diagonal.

USING WINDING STICKS

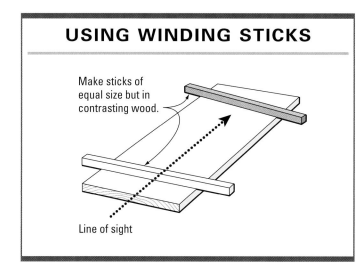

Make sticks of equal size but in contrasting wood.

Line of sight

Handsaws and Chisels

Saw and Chisel Joinery

Installing Hardware

Tuning Chisels

DESPITE THE PROLIFERATION of power tools in today's woodworking shops, handsaws still play an important role. One reason is that power saws—table saws, for example—consume setup time. And the setup time is directly proportional to the complexity of the cut. So it's often more efficient to make the cut with a handsaw. Also, by their nature, handsaws can make cuts that power saws can't. For example, no power tool, saw, or router can duplicate the fine dovetail joinery created with a dovetail saw and a chisel.

Chisels are also among the most essential tools that woodworkers use. They're used for cutting and fitting joints, paring surfaces flush, cutting recesses for locks, hinges, and other hardware, and carving embellishments. Together, handsaws and chisels provide a one-two punch for producing numerous joints common to all types of furniture and casework.

Handsaw Types

Handsaws cut with a long row of teeth. With each stroke of the blade, each tooth, much like a tiny chisel, removes a shaving. As the cut, or kerf, as it is called, is deepened, it has the potential to rub the sides of the sawblade and cause binding. To prevent binding, the teeth of a saw are set; that is, each tooth is slightly bent. Alternate teeth are bent in opposite directions so that the saw will cut a straight path. The resulting

Handsaws are necessary for making cuts that are too cumbersome to make with power saws.

A set of good bench chisels is a central part of any wood-working tool kit.

Handsaws cut with rows of teeth. A close examination reveals a distinct difference between ripsaws and cross-cut saws.

kerf is wider than the body of the blade. However, too much set will cause the blade to wander and make it difficult to saw a straight line. To further reduce binding, better-quality panel saws are tapered from the teeth toward the back of the saw.

A close examination of sawteeth will reveal that their shape is designed specifically for ripping or crosscutting. Ripsaws have large, square-edged teeth for aggressive cutting with the grain. To cleanly sever the tough fibers, crosscut teeth are ground with a bevel that produces a knifelike edge. Crosscut saws are also finer than ripsaws; that is, they have more teeth per inch. This places more teeth in contact with the stock for smooth cutting across the grain.

RIP VS. CROSSCUT TEETH

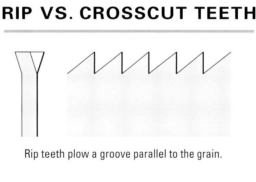

Rip teeth plow a groove parallel to the grain.

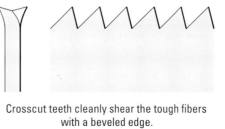

Crosscut teeth cleanly shear the tough fibers with a beveled edge.

Handsaws and Chisels | 97

Panel saws create a wide kerf that is unsuitable for fine joinery.

Dovetail saws are scaled-down backsaws with very fine teeth.

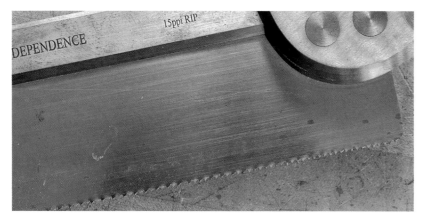

The fine teeth (15 tpi) and slight "set" of this dovetail saw are ideal for sawing precise joints.

A backsaw has fine teeth and a stiff back for sawing fine joinery.

As a handsaw is pushed through the stock, the natural tendency is for the thin steel blade to bow and buckle. Saws are designed to resist buckling in one of several ways. Panel saws incorporate a thick steel body. This helps overcome the problem, but the thick blades of panel saws create a wide kerf unsuitable for fine joinery. Backsaws, as the name implies, have a reinforced back of steel or brass to stiffen the blade during the cut. Large, heavy backsaws are designed for use in miter boxes, while smaller, lighter-weight saws are ideal for sawing tenons. Like backsaws, dovetail saws are also reinforced along the back. However, the steel and "set" of dovetail saws is finer still for producing fine, closely spaced dovetail joints.

In recent years, Japanese handsaws have quickly become a favorite among many woodworkers. Unlike Western saws, which cut on the push stroke, Japanese saws cut on the pull stroke. This design naturally places the blade in tension during each stroke of the saw, which significantly reduces the blade's tendency to buckle and distort. Consequently, the steel of a Japanese saw is thin in comparison to that of Western saws and creates a thin kerf ideal for dovetails and

Japanese saws cut on the pull stroke and create a very fine kerf.

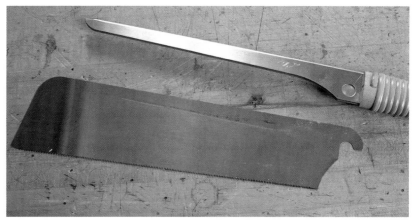

This Japanese saw has a replaceable blade.

The open frame and narrow blade of a coping saw are designed for sawing scrollwork.

A flush-trim saw is "set" only to one side to avoid scratching the surface.

other fine joinery. So that you can avoid sharpening the complex teeth, many Japanese dozuki saws feature a replaceable blade.

Coping saws have a thin, narrow blade for sawing the tight curves in scrollwork. To prevent buckling, the disposable blade is held in tension by the framework of the saw. When replacing the blade, remember that the teeth face the handle, which places the blade in tension during the cut.

The teeth of flush-cutting saws are set in just one direction. This unique design allows you to trim through tenons, pegs, and other joinery flush, without scratching the adjacent surface.

Chisel Types

There are many different types of chisels; the most common is the bench chisel, which features a thin, tapered blade. The sides of the blade are beveled, which allows you to use the chisel in tight, acute corners, like those in dovetails.

Socket chisels provide perfect balance and control for a variety of tasks.

The short length of these socket chisels makes them ideal for chopping the space between dovetails.

Short bench chisels work best for chopping, a process of severing the grain with sharp blows from a mallet. The shorter length, around 9 in. to 10 in., gives greater control when the chisel is positioned vertically and struck with a mallet. To prevent the edge from fracturing from the impact of a mallet, chisels used for chopping should be ground with a steep bevel, typically around 30 degrees.

Paring chisels are longer versions of beveled-edge bench chisels, typically 12 in. or more in length. The greater length provides the leverage that's often required when you're paring and allows you to reach areas that are inaccessible to shorter bench chisels. The bevel angle of a paring chisel is shallow, typically 25 degrees, but sometimes as low as 20 degrees. The shallow bevel gives paring chisels the sharp edge to remove thin, delicate shavings. But don't strike the chisels with a mallet or you may fracture the fragile edge.

CHISEL TYPES

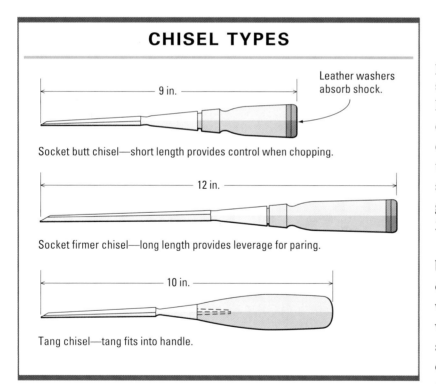

Leather washers absorb shock.

9 in.

Socket butt chisel—short length provides control when chopping.

12 in.

Socket firmer chisel—long length provides leverage for paring.

10 in.

Tang chisel—tang fits into handle.

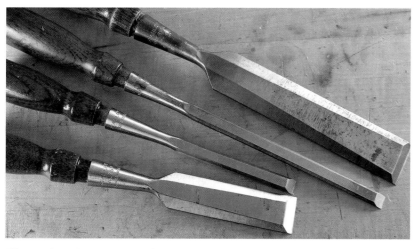

Socket chisels are among my personal favorites.

Short chisels work best for chopping, while longer chisels provide leverage and control when paring.

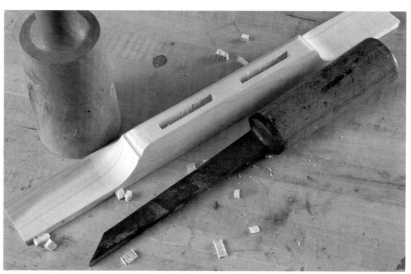

The older-style socket chisels are still my favorites. These strong chisels feature a tapered handle that fits within the forged socket and is driven tight with each use. You can still find these chisels wherever old tools are sold. When shopping for bench chisels, consider the balance and feel of the tool. Chisels with large, heavy handles feel awkward and top-heavy when used for chopping. Try to select a set of chisels that feels comfortable in your hand, and you'll do better work.

Mortise chisels have a thick, narrow blade designed for chopping mortises by hand. The heavy rectangular blade resists twisting when driven by sharp blows of a mallet.

Like Japanese handsaws, Japanese chisels have unique features that distinguish them from the Western varieties. The blades of Japanese chisels are actually laminated from two types of steel. The hard, brittle steel at the cutting edge is supported by a softer, more resilient steel behind the edge. The sides of the blade are sharply beveled for

A mortise chisel has a thick shank for strength.

Japanese chisels feature good balance and hard steel.

reaching into tight areas, and the wood handle is protected by a metal ring for use with a mallet. Additionally, the backs of Japanese chisels are hollow-ground, which makes honing the back easier.

The ends of skew chisels are ground to a 45-degree angle. This creates a sharply pointed edge that allows you to carve into corners that are inaccessible to ordinary bench chisels. Although you can purchase a pair of skew chisels from many woodworking catalogs, it's much less expensive to grind your own from bench chisels.

Carving gouges have a curved face, called a sweep. The curves allow you to carve and shape a variety of embellishments for your work. The degree of curvature, or sweep, is indicated by a number stamped on the shank of the tool. Higher numbers indicate a more pronounced curve. The best carving tools come with a highly polished, mirror finish that allows you to easily hone the tool to a razor edge.

Finally, it's a good idea to purchase a few framing chisels for use in home repair. With tough plastic handles and steel end caps, these tools can take the inevitable punishment that comes when you're working on a deck or other carpentry projects around the house. Save your best chisels for cutting dovetails and other fine work.

A pair of skew chisels will carve into corners.

Chisel Grips

Having good control of a chisel is paramount for creating precise-fitting joinery and other details. Two factors contribute greatly to control of the tool: edge sharpness

Gouges are used for carving.

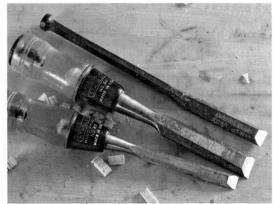

These heavy-duty chisels can take abuse.

and the use of the proper grip. Two surfaces of the chisel merge to create an edge—the bevel and the back. Be aware that most new chisels come with heavy grind marks on the back that must first be polished away. There are several ways to grip a chisel, depending on the type of work you are doing. However, with most every grip, it's important that you position one hand against the workpiece for greatest control. Also, always clamp the work securely in a vise or to the benchtop. Holding the work with one hand while chiseling with the other is an invitation to a nasty cut.

When paring, choose between an underhand grip or an overhand grip. An underhand grip will give you control when removing thin, precise shavings, such as when paring a shoulder of a dovetail or tenon. Grasp the blade of the chisel and grip it with your thumb and index finger. Steady the tool by positioning your index finger against the workpiece.

An overhand grip will give you both power and control for efficiently removing coarse shavings. As you grip the chisel, position the base of your palm against the workpiece. This will provide you with control and leverage. Use a long chisel, and as you push the edge into the stock, use your fist to lever the tool. This technique creates a skewing action that cuts cleaner with less resistance.

When paring a tenon shoulder, place your index finger against the work.

New chisels must be polished on the back.

When paring with the grain, position the base of your palm against the work.

It's easier and more accurate to chisel to a knife line than to a pencil line.

Light cuts across end grain can easily be made without a mallet.

Always position your hand on the work to steady the cut.

Chopping involves holding the chisel perpendicular to the stock and severing the wood fibers. When chopping, always begin by outlining the area with a knife or marking gauge. The chisel edge will easily slip into a thin knife line for great precision. In contrast, a pencil line is thick and leaves no incision to guide the chisel.

When removing small amounts of wood, such as for the lock mortise in the photo above, at right, you can power the tool sufficiently with hand pressure alone—a mallet isn't needed. For greatest control, grip the blade between your thumb and index finger and position your hand against the stock.

For more power, such as for removing the waste when cutting dovetails, drive the chisel with sharp blows of a mallet. If cutting through the stock, cut halfway from both sides. This method ensures accuracy and prevents tearout on the face of the stock. I have a separate set of chisels for chopping

When you grind the sides of your chisels, they can easily trim into the tight spaces of a dovetail.

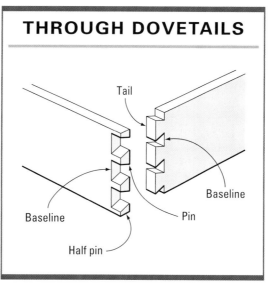

THROUGH DOVETAILS

Tail

Baseline

Baseline

Pin

Half pin

into tight corners, such as during dovetailing; I've ground the sides of the tool to a sharp edge.

Cutting Dovetails

There's probably no better way to develop your skills with handsaws and chisels than by cutting dovetails. Dovetails have been used for centuries as the premier joint for casework and drawer construction. Because of the mechanical interlock between the tails and pins and the numerous long-grain gluing surfaces, no other joint can rival the strength and beauty of the dovetail.

Although myriad dovetail jigs exist, the best method for cutting dovetails is still by hand. Setting up jigs consumes time and material, and besides, no jig can begin to match the beauty of hand-cut dovetails.

Best of all, you'll get the personal satisfaction that comes with skilled handwork. Although seemingly difficult, cutting dovetails is just a matter of sawing and chiseling to a line. The best layout line for dovetailing is an incised line; it's much easier to saw and chisel to an incised line than to one drawn with a pencil. I encourage you to sharpen your tools, review the following pages, and give it a try.

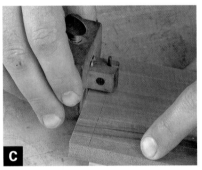

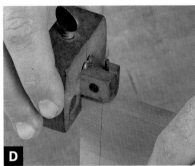

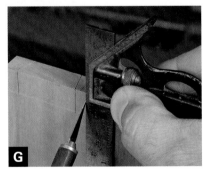

Cutting a Through Dovetail

Begin by selecting soft, straight-grained stock. Although poplar is technically a hardwood, it is physically soft and works well for developing and practicing technique. After milling the stock to size, smooth the surfaces with a bench plane **(A)**. Smoothing the stock after cutting the joint will alter the fit, creating gaps at the baseline of the tails. After smoothing the stock, you're ready for layout. Set a marking gauge to the stock thickness **(B)** and scribe the baseline on both faces of each piece **(C)**. Also scribe the edges of the tail board but not the pin board **(D)**.

The next step is to lay out the pins. First mark the pin spacing with a pencil **(E)**. Then mark the pins with a knife **(F)**. Remember, the pins are sloped on the end grain of the board. Now transfer the layout marks around the corner to the face of the board **(G)**.

Now begin sawing the pins. Secure the board upright in the vise. To start the cut, hold the saw on the corner of the stock adjacent to the layout line. As you pull the saw, guide the blade with your thumb **(H)**. Once the kerf is established, lower the blade and continue sawing to the base-line **(I)**. Once all the sawcuts are made on the pin board, you're ready to remove the waste with a chisel. Position the chisel ⅛ in. away from the baseline and drive it into the wood with sharp blows of a mallet **(J)**. Don't attempt to cut all the way through; instead, cut halfway from each face of the board. Notice how the chisel compresses the wood with a wedging action. Attempting to cut on the baseline initially will force the chisel below the baseline and spoil the joint. However, once most of the waste is cut away, you can accurately cut to the baseline. Position the edge of the chisel in the baseline. Notice how it easily slips into the incision made from the marking gauge. To make the cut, angle the chisel in toward the board so that the base of the joint will be undercut 2 or 3 degrees; this insures a tight fit and yet it doesn't spoil the integrity of the joint **(K)**. Cut halfway, then make a second series of cuts from the opposite face **(L)**.

The next step is to lay out the tails from the pin board. Position the pin board over the tail board, making sure that the narrow part of the pin faces outward. Now trace the pins with a knife to mark the tails **(M)**. Afterwards, transfer the tail layout around the corner to the end of the board, then

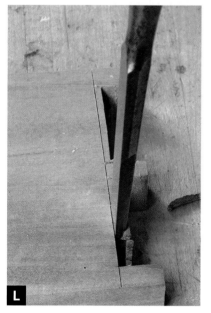

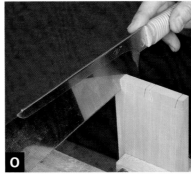

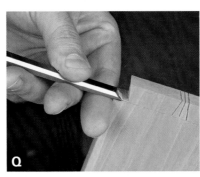

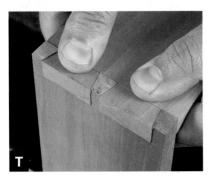

saw the tails **(N)**. To keep the angle consistent, make all the cuts going one way first, then make the cuts angled in the opposite direction **(O)**. Next, saw the area on each corner adjacent to the tails **(P)**. Position the saw slightly above the baseline and finish to the baseline by paring with a chisel **(Q)**. Now chisel to the baseline between the tails **(R)**. To check the fit, gently press the tail board over the pins and note areas that may seem too tight **(S)**. If necessary, pare a shaving off the pin until the joint slips together **(T)**.

Half-Blind Dovetails

Half-blind dovetails can only be viewed from one face of the joint **(A)**. Because one-half of the joint is hidden, the pin board becomes a row of "sockets" to be chiseled out, square and uniform—a great exercise for developing technique with chisels.

Begin by scribing the baselines on the pin board **(B)** and the tail board **(C)**. Use dividers to step off the spacing of the pins **(D)**. Now lay out the pin angles with a bevel or dovetail gauge **(E)** and transfer the layout to the face of the board **(F)**. When building casework, you can save time by clamping mating parts together and transferring the complete pin layout to the opposite case member **(G)**.

(Text continues on p. 110.)

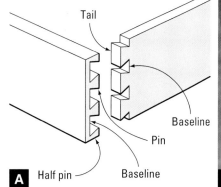

A — Tail / Baseline / Pin / Half pin / Baseline

B

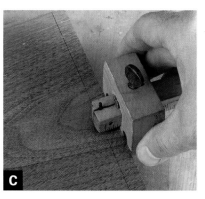

C

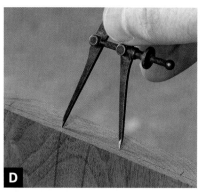

D

E

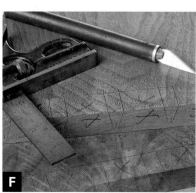

F

G

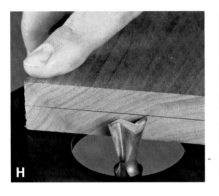

Remove the area between the pins. A router will quickly remove the waste and provide you with the opportunity to chisel the sockets square. Set the router bit depth to the baseline **(H)** and rout the space between the pins **(I)**. Next, chisel the pins and the baseline. Position the chisel in the baseline and cut the baseline to the bottom of the socket **(J)**. As you drive the chisel with a mallet, tilt the handle toward you and undercut the baseline slightly. Now pare the sides of the pins to the baseline **(K)** and clean up the acute, inside corners with a ¼-in. chisel. When complete, the pins should be clean with sharp corners **(L)**.

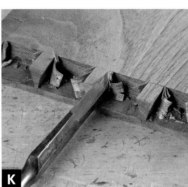

Next, lay out and cut the tails. Position the pin board over the tail board and clamp it into place **(M)**. Carefully trace the pins with a knife **(N)**. Complete the tail layout by transferring the marks around the corner to the end of the board **(O)**.

Now you're ready to saw the tails. Secure the tail board in the vise with the layout close to the jaws to limit vibration as you saw. Make all the cuts in one direction first, then angle the saw in the opposite direction and make the second series of cuts **(P)**. As you saw, space your feet for a broad stance to steady the cut, and use long, smooth strokes **(Q)**. A third series of cuts through the center of the waste piece will make it easier to chisel out the waste between the tails **(R)**. Finally, saw the waste area on either edge of the board **(S)**.

(Text continues on p. 112.)

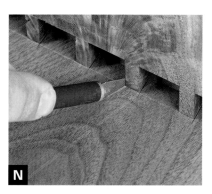

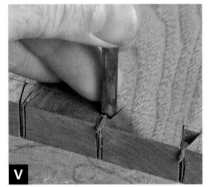

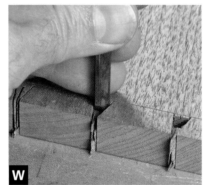

Next, chisel the baseline between the tails. Beginning with the corners, carefully pare down to the incision from the marking gauge **(T)**. Avoid undercutting this area or you'll see a gap when the joint is assembled. Instead, keep the surface square to the face of the board **(U)**. Cut to the baseline between the pins with a narrow chisel **(V)**. After cutting half of the board's thickness, turn the stock over and cut from the baseline on the opposite face **(W)**. To avoid damaging the tails, it's a good idea to bevel the sides of the chisel.

▶ See *"Tuning a Chisel"* on p. 122.

Finally, assemble the joint. If the sawcuts followed the line, a few gentle taps with a mallet will close the joint **(X)**. If the joint seems too snug, don't force it, or you may split one of the boards. Instead, examine each pin closely, and pare a shaving from those that appear to bind until the joint closes completely **(Y)**.

Leg-to-Rail Dovetails

The large, single dovetail that joins a leg to the rail provides yet another good opportunity to practice technique with handsaws and chisels **(A)**. For variation, I'll cut the tail first. The ongoing argument of pins first versus tails first is nonsensical. Both methods work. I suggest that you try both and use whatever way works best for you.

After sizing the stock, begin layout by marking the baseline on both the rail **(B)** and the top of the leg **(C)**. Next, mark the sides of the dovetail on the rail **(D)**. Now you're ready to saw the tail. Begin sawing on the corner. This technique creates less drag on the saw, making it easier to establish an accurate kerf. Use your thumb to guide the blade and pull the saw to start a kerf **(E)**. Now lower the saw and follow the layout line **(F)**. After sawing the tail, carefully pare to the baseline with a sharp chisel **(G)**. Grasp the blade of the chisel between your thumb and index finger and take light, controlled cuts. Steady the chisel by resting your index finger against the stock.

(Text continues on p. 114.)

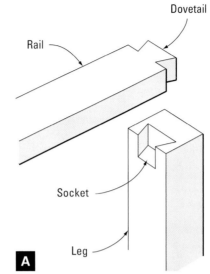

Rail — Dovetail — Socket — Leg

A

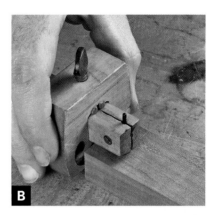

B

C

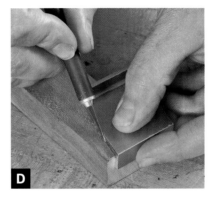

D

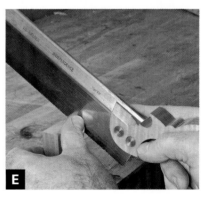

E

F

G

Next, lay out the socket on top of the leg that accepts the tail. As with any layout, accuracy is the key. Secure the leg in a vise and clamp the rail in position over the leg. Use a straightedge to check alignment **(H)**. Now carefully trace the tail with a knife **(I)**.

Next, saw and chisel the socket. Realize that the sawcut is stopped; it doesn't go through the stock. Because the cut is on the corner, it's easier to maneuver the saw if you position the leg in the vise on an angle **(J)**. After sawing, secure the leg to the benchtop with a clamp and chisel out the area between the saw kerfs **(K)**. Now complete the socket at the sharp, acute corners **(L)** and slide the dovetail into place **(M)**. The assembled joint is clean and strong **(N)**.

Mortise-and-Tenon Joint

Like the dovetail, the mortise-and-tenon has a long history in woodworking. Although other contemporary joints exist, such as dowels or wood plates, the mortise-and-tenon remains the choice for strength and longevity. Although it can be easily and efficiently cut with machines, the mortise-and-tenon is a good joint to add to your list of hand-tool skills. Often it's more efficient to cut the joint by hand than to consume time setting up a machine. This is especially true when the joint is complex, such as the compound angled joint found in chairs. As always, accurate layout is a key to success.

Begin by setting a gauge to the tenon length and scribing the shoulder of the tenon **(A)**. Next, mark the location of the mortise **(B)**. The final step in layout is to mark the edges of the stock, which indicates mortise width **(C)** as well as tenon thickness **(D)**. To insure a friction fit between these mating parts, mark them with the same gauge setting to insure accuracy. A sharp marking gauge leaves an incision, whereas the conical points of a mortise gauge tear the wood fibers, leaving an imprecise layout line. Make certain to set the gauge so that the scribe marks equal the thickness of your mortise chisel **(E)**.

[**TIP**] **With any mortise-and-tenon joint, always cut the mortise first, then cut the tenon to fit. It's always easier and much more precise to adjust the fit by shaving the tenon rather than enlarging the mortise.**

Before mortising, stick a small strip of tape to the chisel to indicate the mortise depth **(F)**. Begin mortising approximately ¼ in. from one end of the mortise and work toward the opposite end. Hold the chisel vertically with the bevel turned inward,

(Text continues on p. 116.)

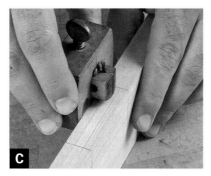

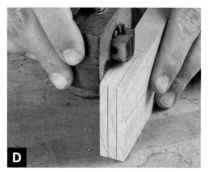

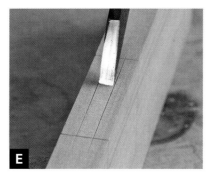

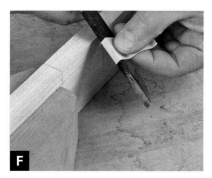

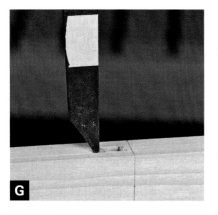

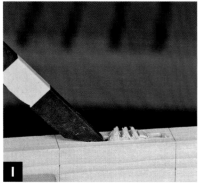

and drive the tool with sharp blows of a mallet **(G)**. Lever the chisel against the bevel to remove the chips **(H)** and work your way to the opposite end of the mortise **(I)**. After removing most of the waste, position the back of the chisel at the end of the mortise and chop the end wall square **(J)**.

With the mortise complete, saw the tenon. The surface of a tenon is broad, so for accuracy, it's best to saw from opposite directions and form two kerfs that meet in the middle. Position the workpiece in the vise at an angle and begin sawing into the corner **(K)**. Next, turn the workpiece in the vise and saw the opposite corner until the kerfs meet **(L)**. Now, make a third cut that completely joins the two **(M)**. When making the third cut, saw within $\frac{1}{32}$ in. of the shoulder of the tenon. For the final sawcut, you'll need a bench hook.

▶ See *"Bench Accessories"* on p. 47.

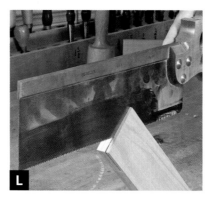

Position the workpiece against the stop on the bench hook and align the saw approximately ¹⁄₁₆ in. from the tenon shoulder. Now saw down to the face of the tenon **(N)**.

Next, use a wide chisel to incise the shoulder of the tenon. As you work across the width of the tenon, guide each cut by slipping the chisel edge into the incised layout line **(O)**. The final step is to smooth away the saw marks with a rabbet or shoulder plane **(P)**. Test the tenon for a snug but not tight fit within the mortise; **(Q)** the joint should assemble with firm hand pressure **(R)**.

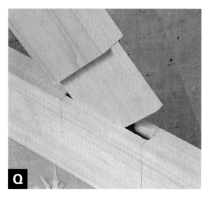

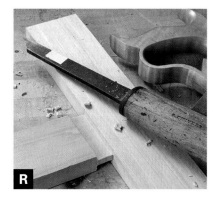

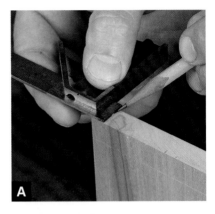

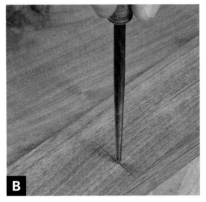

Cutting a Lock Mortise

The practice of installing locks on furniture goes back hundreds of years, to when people secured important documents, jewelry, and spices inside drawers and behind doors of fine casework. Although most people today don't keep spices under lock and key, a lock still adds a touch of class to a fine piece of furniture. And installing a lock is a good exercise in learning to use a chisel.

Begin by measuring for the location of the keyhole (**A**). Mark the spot with an awl to prevent the drill from wandering (**B**). Now, drill a row of two or three holes to form an elongated hole for the key (**C**). Position the lock over the keyhole and lay out a shallow mortise for the body of the lock (**D**). Now select a wide chisel for cutting the mortise. Hold the chisel vertically with the edge perpendicular to the grain, and chop a row of incisions (**E**). These cuts will sever the grain and make paring easier and more efficient. Pare away the incised wood and repeat the process until the lock body fits the mortise.

Next, cut a shallow mortise for the backplate of the lock. Position the lock body in the mortise and align it with the keyhole. Then outline the perimeter of the backplate with a knife **(F)**. Now incise the wood to the outline and pare to a depth of the backplate thickness **(G)**. Check the fit, and, if necessary, pare a shaving or two until the backplate is flush with the wood surface, and then install the lock in the mortise **(H)**.

The last step of the process is to cut a mortise in the cabinet for the bolt of the lock. Begin by extending the lock bolt and marking the bolt location on the face of the cabinet **(I)**. Now transfer the layout lines into the doorjamb. Drill a row of shallow holes and complete the bolt mortise by squaring the perimeter with chisels. A narrow, ⅛-in. wide chisel will cleanly square the ends of the mortise **(J)**. Make sure the bolt operates smoothly **(K)**.

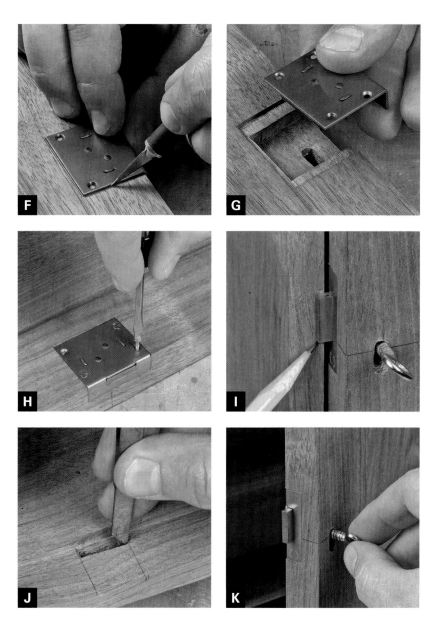

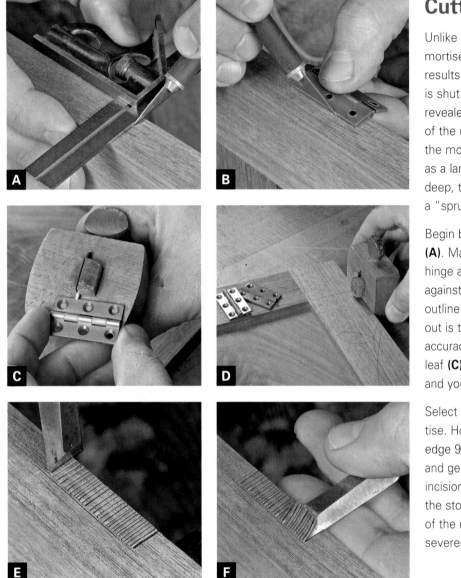

Cutting a Hinge Mortise

Unlike surface-mount hinges, butt hinges are mortised into the door stile and the cabinet. The results yield an uncluttered look; when the door is shut the barrel of the hinge is all that is revealed. For the door to fit properly, the depth of the mortise must equal the leaf thickness. If the mortise is too shallow, the reveal will appear as a large, unsightly gap. If the mortise is too deep, the door may bind on the cabinet, creating a "sprung" door that will not stay shut.

Begin by marking the location of the hinge ends **(A)**. Mark the width of the mortise using the hinge as a gauge. Position the barrel of the hinge against the edge of the door stile and mark the outline with a knife **(B)**. The final step in the lay-out is to mark the mortise depth. For the greatest accuracy, set a marking gauge from the hinge leaf **(C)**. Now scribe the depth on the door **(D)**, and you're ready to cut the hinge mortise.

Select a chisel that is slightly wider than the mortise. Hold the chisel vertically and position the edge 90 degrees to the wood grain. Use a mallet and gently tap the chisel to cut a row of shallow incisions **(E)**. Now position the chisel parallel to the stock and pare across the grain to the depth of the mortise **(F)**. Because the grain was first severed, it will pare easily. Next, place the hinge in the mortise and check the depth **(G)**. If necessary, pare a bit more until the hinge is flush with the surrounding wood.

The mortises on the cabinet will be marked directly from the door. First mark the location for the hinge screws **(H)**. Then drill the screw holes **(I)** and mount the hinge to the door **(J)**. (Drive steel screws first to form tight pilot holes; otherwise the brass screws might break.)

Position the door in the opening and mark the hinge locations on the cabinet **(K)**. Use wooden shims to hold the door in position and provide an even space around the perimeter of the door. Now repeat the mortising procedure on the cabinet and hang the door **(L)**.

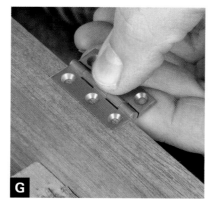

G

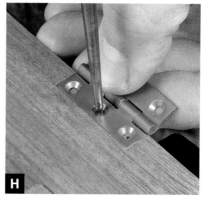

H

I

J

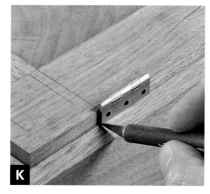

K

L

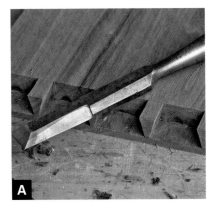

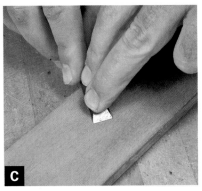

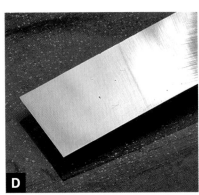

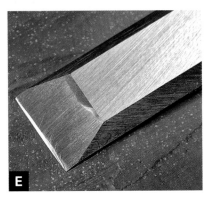

Techniques for Tuning Chisels

Chisels are used for a variety of tasks, from chopping the waste between the pins of a dovetail to paring a thin shaving when fitting a hinge **(A)**. Although chisels are simple tools, there are still a few steps you can take to make them work more effectively. As with all edge tools, for the greatest control a chisel must be sharp. And remember that the back and bevel merge to create the cutting edge. If you focus your sharpening efforts on the bevel alone, you've only sharpened one-half of the edge.

Most new woodworking chisels come from the factory with heavy grinding marks that are incompatible with a sharp edge **(B)**. As the grinding marks reach the cutting edge, they form tiny serrations that lift and tear the wood. Rust pits also form small nicks and serrations at the cutting edge. So if you're shopping for old chisels, avoid those with excessive rust.

To prepare a chisel for sharpening, begin by rubbing the back of the tool across a coarse sharpening stone to remove the grind marks or discoloration from surface rust **(C)**. This process will flatten the back as you polish the steel with progressively finer stones. After its polishing with the last benchstone, the surface should have a mirror finish **(D)**. Next, grind the bevel and hone it as you would any edge tool **(E)**.

▶ See *"Sharpening a Chisel"* on p. 175.

The next step in the tuning process is to grind the side bevels of the chisel **(F)**. A sharp, almost knifelike edge along the sides will allow you to cut between dovetails without crushing the adjacent surfaces. As you grind the edge, use a light touch and follow the angle of the original bevel. There's no need to further hone or polish the sides.

Next, turn your attention to the handle. When you're driving a chisel with a mallet, it's important that each blow land squarely on the end of the handle. Unfortunately, many new chisels come equipped with handles that have rounded ends **(G)**; when the chisel is struck, the mallet tends to glance off. The solution is to saw a small portion off the end of the handle **(H)**. Then, smooth away the sawmarks and chamfer the edges slightly with a file **(I)**. When finished, the end of the handle should resemble those found on older socket chisels **(J)**. Once you've spent a few minutes tuning the chisels in your kit, you'll quickly see improvements in your joinery **(K)**.

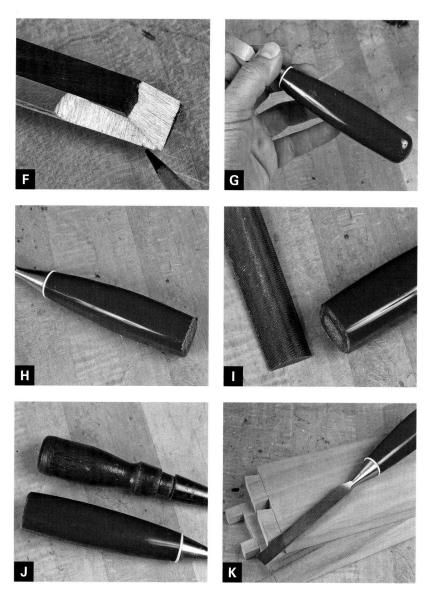

Planes and Planing

Planing Techniques

Planing Project Parts

Planing Shapes

Tuning Planes

PLANES ARE UNIQUE AMONG woodworking tools; few tools are quite as useful and yet such a pleasure to use. No other tools have the finesse to remove long, consistently thin shavings from highly figured hardwoods. Their pleasant, wispy sound as they pass over the wood and the glistening, smooth surface they leave behind combine to create one of the most enjoyable of woodworking experiences.

Bench Plane Anatomy

As the edge of a plane iron enters the wood, it encounters a considerable amount of resistance. A well-constructed plane, when tuned and sharpened, can effectively overcome the resistance as it lifts and curls a shaving. But as the edge of the iron enters the wood and lifts a shaving, the natural tendency is for the blade to continue below the wood surface, producing tearout. To prevent tearout, the sole of the plane, just ahead of the mouth, places pressure on the stock. At the same time, the cap iron breaks and curls the shaving upward, effectively eliminating tearout. However, if the plane is poor quality, needs adjustments, or if the blade is simply dull, the cutting edge will vibrate or "chatter," and the plane may clog with shavings or skip and hop across the board. Several factors contribute to successful planing. Simply stated, planing requires a thick, sharp iron that is firmly bedded to a heavy plane. Let's take a closer look at the dynamics of planing.

If the wood you're working is soft pine and you're planing in the direction of the grain, the resistance is minimal. However, if the wood is dense and figured, such as with tiger maple, the resistance is substantial. When the plane is tuned for optimum performance, the edge is supported by the other parts of the plane, such as the cap iron, frog, and main body. These assorted parts, plus the mass of the plane, work to absorb and distribute the force of planing. In fact, with everything else equal, a heavy plane performs better than a light one. A small mouth is important, too. Bench planes have an adjustable frog to control the size of the mouth and adjust for the thickness of the shaving desired. Other types of planes, such as block planes, have an adjustable toe that serves the same purpose.

Old Versus New Planes

Years ago, plane manufacturers such as Stanley, Sargent®, and Ohio Tool® made planes for nearly every conceivable purpose. There were planes for truing and sizing

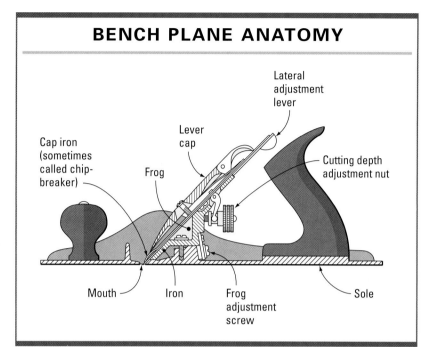

BENCH PLANE ANATOMY

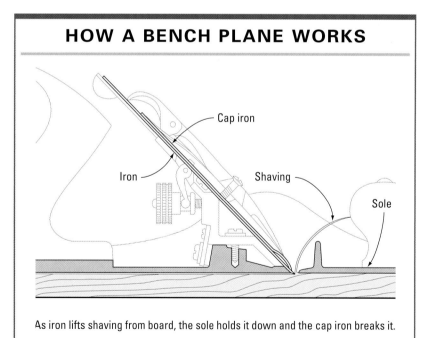

HOW A BENCH PLANE WORKS

As iron lifts shaving from board, the sole holds it down and the cap iron breaks it.

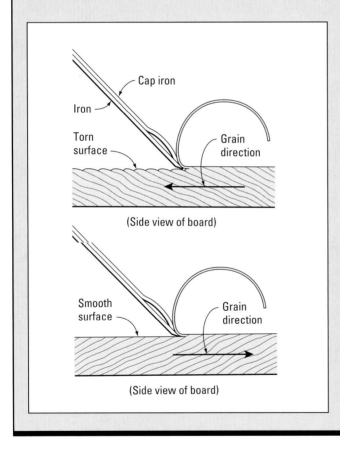

► HOW GRAIN DIRECTION AFFECTS PLANING

Effective planing requires more than the skill of using a plane; it also requires that you read the wood. No matter how sharp the plane is, if you plane against the grain (top), the wood fibers will tear out. Planing with the grain (bottom) will yield perfect results.

Cap iron

Iron

Torn surface

Grain direction

(Side view of board)

Smooth surface

Grain direction

(Side view of board)

stock, planes for smoothing the wood's surface, planes for cutting and fitting joints, and even planes for shaping curves and molding profiles. Through the years following World War II, a number of factors caused the quality and variety of planes produced to drop significantly. Fortunately, many fine old planes survive and can be found at flea markets, tool auctions, and estate sales. With cleaning, tuning, and perhaps a new iron, these beautiful old planes can be made to perform as well, or perhaps even better, than when new.

Today there are new sources for high-quality planes, too. As the popularity of woodworking has expanded, entrepreneurs such as Tom Lie-Nielsen have risen to the demand for high-performance planes. Based upon designs from Stanley, Lie-Nielsen® planes include dramatic improvements, such as thicker castings and irons and more precise fittings. Additionally, cryogenics has significantly improved the edge retention of the steels used for blades since the early days of Stanley. Undoubtedly, whether using a refurbished classic or a newer, high-tech model, woodworkers today have access to some of the finest planes ever produced.

Bench Plane Sizes

Bench planes are the workhorses among planes. When I need to carefully fit a drawer, flatten a wide board, or smooth away mill marks after sizing stock on machines, I reach for a bench plane. Bench planes range in size from the diminutive No. 1 to the huge No. 8. Nos. 1 through 4 are referred to as smooth planes, intended for smoothing the rough surface created by larger planes, or, more

Bench planes are among the most common and useful handplanes.

The most popular bench plane is the No. 4 smoothing plane.

The No. 4½ is better suited for smoothing wide panels.

Long bench planes are best for flattening rough stock and straightening edges.

commonly today, the milled, textured surface created by the table saw, jointer, and planer. The tiny No. 1, although a working plane, is more a tool collector's novelty and is too small to be practical. At 9 in. long and 3¾ lbs., the No. 4 is the most popular size for smoothing. The No. 4½ is slightly longer, wider, and heavier than a No. 4 and is a favorite for smoothing large panels for tabletops and sides for casework.

The larger bench planes, Nos. 5 through 8, are designed for truing the faces and edges of stock. The long, flat soles of these planes will span the low areas to remove only the high spots and create a flat edge or surface.

Bedrock Planes

In the years following the Civil War, plane makers such as Leonard Bailey were devising improved metal planes to replace the

► SANDING VERSUS PLANING

I remember working on a project in high school shop many years ago. I was aggressively sanding a joint in an attempt to level the mating surfaces. The instructor stepped up behind me with a sharp plane and leveled the joint in what seemed like an instant. Even better, the surface was smooth—I mean really smooth. Ever since, I've been hooked on planes.

Sandpaper smoothes the wood surface by abrading, and it leaves behind telltale scratches. Machines such as planers and jointers smooth and flatten stock with a series of small cuts as the cutterhead spins. This machining process produces a series of small but distinctive ridges. Planes, however, slice through the wood fibers and remove one long, continuous shaving, creating an incredibly smooth, polished surface.

wooden planes that were common at the time. With lateral adjustment levers, screw feed depth adjusters, and iron soles (which, unlike wooden soles, didn't expand and contract with the weather), metal planes quickly caught on. Leonard Bailey was bought out by Stanley and the Bailey-type planes, essentially unchanged for the past 100 years, are still popular today.

In the early 20th century, Stanley Tools developed an improved version of its popular "Bailey" bench planes. Labeled "Bedrock," the new, more expensive design featured a much-improved frog that gave much greater support to the iron. Compared to the earlier design, the frog in the Bedrock design was firmly bedded to a broad machined surface in the body of the plane, hence the name Bedrock. Additionally, grooves milled along the edges of the frog kept it aligned with the mouth of the plane. Furthermore, adjustments to the frog position could be made without removing the lever cap and iron

Bedrock planes can be identified by the "60" prefix. The plane at left is a standard No. 6; at right is a Bedrock No. 606.

The frog of "Bedrock"-style planes is a much-improved design over "Bailey" planes.

from the plane. Rather than screws to secure the frog to the plane body, Bedrock planes used pins that are dimpled to engage with screws from the back of the frog. Stanley continued to sell its less costly Bailey-type planes, so to make it easier for the public to distinguish between the two, it modeled square sides into the body of the Bedrock planes. Although the frog design is a dramatic improvement, Bedrock planes still suffered from the thin cutting iron that was, and is still, provided with the Bailey-type planes. The idea was that a thin iron was easier to sharpen. Remember, too, that these planes were primarily intended for use on

straight-grain softwoods rather than figured hardwoods. Even so, the long-discontinued Bedrock planes sell for many times more than when they were new. Woodworkers search for usable examples that they can put to work, while tool collectors desire the pristine planes, preferably with the original box. When cleaned, restored and fitted with a new, thicker iron (see p. 162), Bedrock planes perform quite well and considerably better than Bailey-type planes.

As the popularity of woodworking has risen, so has the demand for high-quality hand tools. Several companies are now producing new lines, including layout tools, edge tools, and planes, to meet that demand.

Lie-Nielsen Toolworks

Lie-Nielsen produced its first plane in the early 1980s. Based upon the Bedrock designs, Lie-Nielsen bench planes feature a mating, machined fit between the plane body and the frog. Lie-Nielsen also improved the Bedrock design by adding thicker castings, thicker irons, and cryogenically treated double-

A machined groove in the Bedrock frog keeps it aligned with the sole.

A Bedrock plane is easily distinguished by its square sides.

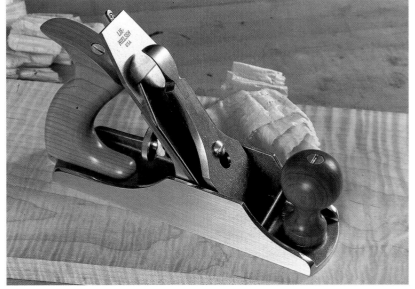

This Lie-Nielsen smooth plane is an improved version of Bedrock designs.

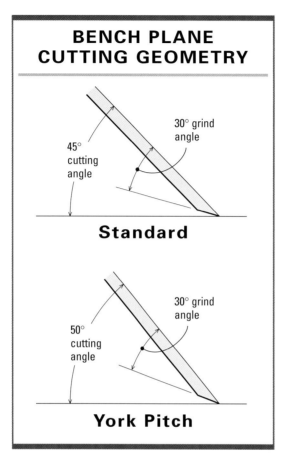

BENCH PLANE CUTTING GEOMETRY

45° cutting angle

30° grind angle

Standard

50° cutting angle

30° grind angle

York Pitch

tempered steel. Additionally, you can also purchase Lie-Nielsen bench planes with a high-angle frog. On a standard bench plane, the frog supports the iron at a 45-degree angle. However, the high-angle frog, often referred to as a York pitch, supports the iron at a 50-degree angle. (1302) The steeper pitch provides a noticeable improvement in cutting quality over the typical Bailey-style plane, particularly in highly figured hardwoods. Undoubtedly, Lie-Nielsen planes are among the finest available to woodworkers today.

Block Planes

The block plane is another tool that should be part of every woodworking tool kit. Because of their small size, block planes are useful for a wide variety of planing tasks. I have several types of block planes, and I use them frequently for everything from surfacing small sheets of freshly bandsawn veneer to trimming and fitting doors and drawers.

In some ways you can think of block planes as scaled-down bench planes. Both are used for smoothing and both typically

York pitch planes perform better on highly figured woods.

Block planes are small enough for one-handed use.

have a 45-degree cutting angle. But there are also distinct differences that affect their use. For example, as mentioned earlier, you can purchase a high-angle bench plane for smoothing difficult grain, while low-angle block planes are available for trimming end grain. Additionally, block planes don't have an adjustable frog, but better ones have an adjustable mouth at the toe of the plane that serves the same purpose. Pressure may be applied to the cap with a lever or spinwheel. Some old Stanley block planes feature an unusual "knuckle-joint" lever cap. There are other differences, too, between block planes and bench planes. Block planes typically

The "mouth" on a better-quality block plane is adjustable, allowing for finer control and cutting.

This small block plane uses a spin-wheel cap to apply pressure to the cutting iron.

This old low-angle block plane features an adjustable mouth and a knuckle-joint lever cap.

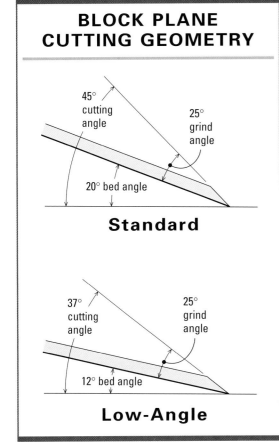

BLOCK PLANE CUTTING GEOMETRY

45° cutting angle

25° grind angle

20° bed angle

Standard

37° cutting angle

25° grind angle

12° bed angle

Low-Angle

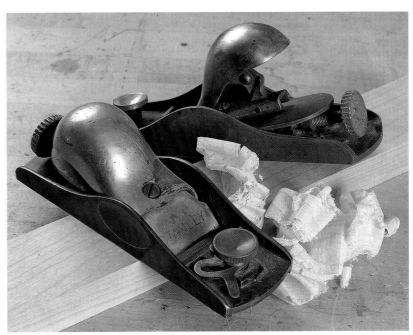

The cutting iron of a block plane is bedded in the plane body with the bevel up.

This rabbet block plane is ideal for trimming tenons.

You can smooth and trim the toughest end grain with this large block plane.

This edge-trim block plane features a 90-degree fence.

don't have a cap iron, and the iron itself is bedded bevel up. These differences result in a small, easily maneuverable plane that is useful for a variety of planing tasks, yet most block planes are not nearly as effective at planing difficult grain as the best bench planes.

At first glance, the tool at the left, in the top left photo above, looks like a bench plane. However, because it lacks a cap iron but has the iron bedded bevel up and a low angle, it's really a large block plane. As on most block planes, the frog is cast as part of the plane body, and mouth adjustments are made by sliding the casting at the toe. This

plane is a great choice anytime you need to smooth large end-grain surfaces.

The rabbeting block plane features an open mouth on both sides of the plane. This unusual block plane makes fitting large tenons a snap. And the knurled thumbwheel at the back of the plane makes it easy to adjust the cutting depth.

A rather unique tool is the edge-trim block plane. With an integral fence, this plane makes it easy to trim and square 90-degree edges. The skewed blade makes clean cuts, especially on end grain.

Scraping Planes and Scrapers

Although not usually considered planes, scrapers are tools for smoothing wood that work much like high-angle planes. But unlike planes, scrapers cut with a tiny burr. It's the burr that makes scrapers special; it limits the cutting depth to reduce tearout to a bare minimum on even the most difficult woods. It is also the burr that makes scrapers so difficult to sharpen. (See "Sharpening Hand Tools," p. 168.) But once you understand the steps involved, it is really quite easy. Using a sharp scraper is quite an experience—it will dramatically reduce the time that you spend sanding. Remember, sanding removes wood by abrading. To smooth effectively, sanding must be done with progressively smaller grit sizes. However, a sharp scraper removes shavings, not dust, so it's much more efficient than sandpaper.

There are several types of scrapers to choose from. The humble appearance of the cabinet scraper belies its capabilities. It's effective for a variety of scraping jobs, from smoothing a difficult area in a door panel to tight curves in a chair leg. To use, simply flex it between the fingers and thumbs.

For scraping the framework of doors and casework, my favorite is the small scraping plane. Its thick blade limits chatter, and the short sole keeps surfaces flat yet allows for easy maneuverability.

When you're scraping tabletops, case sides, and other large panels, the large scraping plane is the tool of choice. The large sole and wide blade eliminate dips and irregularities during scraping. The handle and knob provide a positive grip and allow you to use long, powerful strokes.

This small scraping plane will level joints on the most difficult woods with no tearout.

A scraping plane will smooth large, highly figured panels with no tearout.

A sharp cabinet scraper will quickly remove tearout and other surface defects.

► USING A CABINET SCRAPER

The cabinet scraper is a deceptively simple tool. But getting the hang of it takes some patience, and it also needs to be properly sharpened.

To use a cabinet scraper effectively, first secure the work to the bench.

Grasp the scraper with both hands and flex the steel with your thumbs. Tilt the scraper slightly toward the direction in which you're cutting.

When the scraper is sharp, it will produce thin, fluffy shavings.

To remove small areas of tearout, scrape a larger area around the defect to avoid creating a depression.

This large, heavy shoulder plane will remove precise shavings.

The sides of a shoulder plane should be absolutely 90 degrees to the sole.

Planes for Fitting Joints

Almost every woodworking project you'll undertake will have at least one joint, and there are a variety of planes for cutting and fitting joints. No shop should be without at least one shoulder plane; my own shop has several. Often referred to in woodworking catalogs as a rabbet plane, the shoulder plane is designed for trimming cheeks and shoulders of tenons for a precise fit with the mating mortise. While most rabbet planes take a coarse shaving, a true shoulder plane has a fine mouth and a low cutting angle for precise shavings across the end grain of shoulders. Once you've learned to tune and use this tool, you'll find the clean, controlled cuts you'll get with a shoulder plane addictive. A large, heavy shoulder plane is the best choice for trimming the end grain of broad tenons. The best ones feature sides that are ground perfectly square to the sole. This ensures that the tenon shoulder is trimmed square to the face of the tenon. Smaller shoulder planes are useful, too, for trimming in close areas where a larger plane may not fit. Some shoulder planes feature a removable toe that converts the tool into a chisel plane for trimming into corners.

The nose of this shoulder plane slips off to convert the tool to a chisel plane.

Small shoulder planes will cut into tight areas where a larger plane won't fit.

Rabbet planes feature nickers, which score the fibers ahead of the cutting iron when you're planing cross-grain rabbets.

This rabbet plane features a fence and depth stop to control the size of the rabbet.

Rabbet Planes

Like shoulder planes, rabbet planes feature a mouth open at the sides for trimming into corners. However, most rabbet planes are designed specifically for cutting, not just trimming rabbets. The most elaborate rabbet planes feature nickers, fences, and depth stops. The nickers, or spurs, sever the fibers ahead of the iron to prevent tearout when you're cutting across the grain. Fences and stops control the size of the finished rabbet. The fence rides the edge of the stock to control the rabbet width. Successive cuts are taken, and eventually the depth stop rubs the surface of the board and limits the cutting depth.

Flip through the pages of a woodworking tool catalog, and you'll see a number of rabbet plane designs from which to choose. The rabbet block plane features a skewed iron for cleaner cuts. It also has a removable side plate on the main body so that this unique tool can serve double duty as a block plane when it's not cutting rabbets. A small fence attaches to the plane to guide the cut during

A skewed iron on this little rabbet plane makes for cleaner cuts.

This large "jack-rabbet" plane will smooth large rabbets with ease.

▶ SPECIALTY PLANES

In the heyday of plane-making, there were dozens of specialty planes manufactured, each having a very unique and specific function. Here are two examples. A quite odd-looking tool is the matched tongue-and-groove plane. Equipped with two handles and two irons, this plane can be pushed in one direction for cutting a groove and in the opposite direction to cut the mating tongue. A fence cast into the body of the plane guides each cut, and integral stops control the cutting depth. The swing-fence match plane also cuts a matching tongue-and-groove joint. Fitted with two separate irons to cut a tongue, the offset fence pivots to cover one iron and cut the corresponding groove.

You can work this tongue-and-groove plane in either direction.

This tongue-and-groove plane features an offset swinging fence.

rabbeting. The jack-rabbet is essentially a bench plane with open sides. Its large size, extra mass, and handles will allow you to cut or smooth large rabbets with ease. The handles on this unusual plane tilt so that you can plane close to corners without banging your knuckles.

Spokeshaves

Spokeshaves have been around for centuries. They're essentially a short-soled plane with handles on either side, and they're the tool I reach for when smoothing and fairing band-sawn curves. Like many planes, spokeshaves are available in both wood and metal versions. With its low cutting angle, the wooden version is a true shave and is the best choice for working the green riven spindles of a

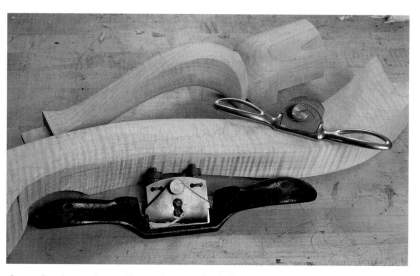

A spokeshave is really a short-soled plane that is ideal for smoothing curves.

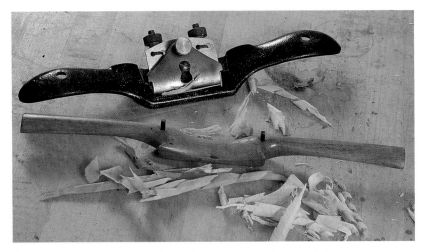

Spokeshaves are available with either wood or metal bodies.

This small spokeshave will smooth tight contours.

The flexible steel sole of a compass plane will flex to smooth contoured surfaces.

Windsor chair. For most woodworking, however, I prefer a metal spokeshave. Its higher cutting angle and greater mass combine to make smoother cuts on hard, dried stock. Like most metal planes, metal spokeshaves have thumbwheels that make cutting depth adjustments quick and positive. Years ago, metal spokeshaves were made in numerous sizes and varieties. The soles were available round or flat, concave or convex. Sometimes two profiles were on the same shave to eliminate switching between tools. Because there were so many manufactured, old spokeshaves are still widely available today. And a number of various types of spokeshaves are still being produced today. Despite all the varieties, you're likely to find that the ordinary flat-bottomed shave is the most useful.

Compass Planes

With their heavy iron bodies and milled soles, most planes are designed to create flat, true surfaces. But the compass plane has a thin, flexible steel sole that allows it to flex for smoothing curved surfaces. An adjustment knob on top of the plane pushes or pulls the sole into a convex or concave contour. While the compass plane is a useful tool, it works best on arcs or circles, rather than free-form curves.

Molding Planes

In the years before power tools, cabinetmakers fashioned moldings for their furniture using specialized molding planes. Each molding profile required a separate plane that was pushed repeatedly over the stock until the full depth of the profile was

► WHY USE PLANES FOR SHAPING?

Despite the proliferation of routers, jigs, and templates, planes designed specifically for shaping are some of the most useful you can own. Molding planes, spokeshaves, compass planes, and scratch beaders were the routers and shapers of their day. But why use them today? Because fine furniture has many subtle details that machines can't create. The sensuous curves of a cabriole leg are still best refined with a sharp spokeshave. And only a scratch stock can shape a quirk bead on the compound curve of a chair's crest rail. These quiet tools are much more enjoyable, too, than any power tool.

Wooden molding planes are still useful for shaping moldings.

This wooden plane will shape a beautiful ogee profile.

The sole of a molding plane has the negative shape of the profile it creates.

reached. Large, complex moldings could also be cut with a single plane, but large molding planes were expensive and difficult to push. Instead, a series of smaller planes were often used to shape wide moldings. To determine the shape of a molding created by a specific plane, turn it over and examine the sole.

The sole of the molding plane, along with the iron, has a profile that's the opposite of the molding profile it creates. A slender wooden wedge locks the iron firmly into the body of the plane. As the shaving is produced, it is pushed out the side of the plane.

Molding planes use a simple wooden wedge to apply pressure to the cutting iron.

This sidebead plane shapes a bead along the edge of a board.

Hollow and round planes like these were numbered.

The maker's mark indicates that this is a matching pair.

Profiles of molding planes are familiar— ogee, ovolo, quirk bead—to name a few. The most versatile profiles, hollows and rounds, are still common at flea markets and tool auctions everywhere. Hollows have a concave sole that produces a convex profile. Rounds shape a matching concave profile. Hollows and rounds were once produced in matching numbered pairs and full sets of up to 24 sizes. If you are patient, it is not diffi-

cult to find matching pairs today, or even partial or complete sets. And you're sure to find plenty of uses for them. Besides shaping large complex moldings that otherwise require a heavy, industrial shaper, hollows and rounds can also be used for shaping and sculpting curves.

When considering molding planes for purchase, look for those that are complete with iron and wedge. It's difficult and very

Complete matching sets of hollow and round planes can still be found and are quite useful.

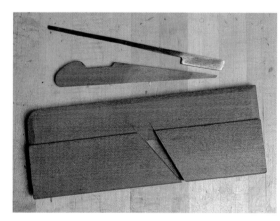

Avoid buying molding planes that are missing the cutting iron or wedge.

Look for wooden planes that have little wear on the sole, as indicated by a tight mouth.

time-consuming to fashion a new wedge and/or iron to fit an old plane. Avoid planes with excessive wear on the soles or split ends due to contact with moisture. The mouth should be tight in order to effectively break the shaving as it is lifted by the iron. If the sole shows signs of excessive wear, or was reshaped due to wear, the mouth will be large, and the plane will have a greater tendency to tear the wood. The soles of better-quality planes were carefully fitted with dense boxwood strips along wear points to extend their working life. Wooden molding planes were produced in mass quantity by a number of companies up until the late 19th century. Many survive in excellent condition,

► THE ESSENTIAL PLANE KIT

If you are new to woodworking and have not begun to acquire planes, you may be wondering where to start. A bench plane is a good first choice. Once you learn to use this essential tool, you'll reach for it often for smoothing panels, fitting drawers, and leveling joints at assemblies.

Woodworkers disagree on the optimum size for a smooth plane; no doubt that's why there have always been a number of different sizes from which to choose. I feel that the No. 4 is the best choice for the first smoother. The plane has enough weight to give it power through the stroke, yet it has excellent balance and maneuverability. Close cousins to the No. 4, the No. 3 is just slightly too small and light, and the No. 4½ is a bit heavy.

The bench, block, and shoulder planes (right to left) are essential for any furniture craftsman.

Later on, consider adding a No. 6 or No. 7 to your tool kit. Either of these bench planes will allow you to flatten wide boards that exceed the width of your jointer. They're also useful for fine-tuning the edges of stock before glue-up.

The second plane on my list of essential hand tools would be a block plane. This small, one-handed plane can be used for a number of light tasks, from smoothing the end grain of a small tabletop to fitting small drawers to light chamfering and shaping. Look for a low-angle plane with an adjustable mouth.

The third and final plane on my essentials list would be a shoulder plane. Unlike the previous two planes on the list, this tool has open sides at the mouth so it can trim into corners. Shoulder planes are the best tools for trimming and fitting tenons to their mortises. Although I prefer a large shoulder plane, one of the fine-quality smaller planes is a good choice.

Determining what to add to your list of planes will become easier once you've mastered using these three. As your woodworking knowledge and skills increase, you'll develop your own working style, and you can add planes as the need for them arises.

This "boxed" plane has a boxwood insert
at the wear point, which significantly
extends its working life.

This metal scratch stock has interchangeable cutters of various
profiles.

show little evidence of wear, and can be had
at a reasonable price, so it just isn't worth the
time and trouble to rework those in poor
condition.

Scratch Stock

Part molding plane and part scraper, scratch
stock are best defined as small, high-angle
molding planes. They excel at effectively
shaping small, simple profiles, especially quirk
beads, on curved surfaces and difficult woods
on which a router may not be effective.

Scratch stock cut like a scraper—with a
small burr. However, the edge is profiled like
that of a molding plane. You can purchase a
metal-frame scratch stock along with several
profiled cutters, or you can make one of your
own with an old marking gauge or even a
scrap wood block. If you need to shape profiles along the edge of concave curves, shape
a curve on the face of the gauge. The steel

for a scratch stock cutter should be hard
enough to hold an edge, yet soft enough to
take a burr. Old handsaw blades are ideal.
You can shape the profile on the cutter with
small files.

A scratch stock can
easily be made
from a wood block
and a piece of an
old cabinet scraper.

This Stanley No. 45 will adjust to cut beads or rabbets.

Stanley Models 45 and 55

Throughout the late 19th and early 20th centuries, Stanley Tools was busy devising its own metal versions of wooden planes. The No. 45 and later the No. 55 were the company's metal versions of molding planes. The No. 45 was marketed as a combination plane—able to cut a variety of symmetrical profiles, such as rabbets, dadoes, and quirk beads. Soon to follow was a more complex version of the plane, labeled No. 55. Stanley's No. 55 could also cut asymmetrical profiles such as the ogee and thumbnail. The complete 55 plane, with all its cutters, fences, and accessories, was boldly advertised as a "planing mill within itself."

However, although both of these planes are functioning tools, the No. 55 is quite awkward to use. And with the many parts included with these planes, both can require a bit of setup time, especially for the more complex molding profiles.

Undoubtedly, many earlier craftsmen felt the same; it's not uncommon to find these old planes in their original boxes with little or no signs of wear.

The Stanley No. 55 will cut a number of different molding profiles

Basic Planing Mechanics

Despite the wide variety of hand planes, the techniques for using them are often very similar. So once you've mastered the basic technique, you are well on your way to performing a variety of planing tasks. A good place to begin is by using a No. 4 bench plane to smooth the edge of a board. First make certain that the plane is sharp and well-tuned (see p. 158).

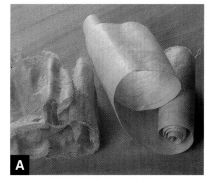

When I reach for a plane, I usually check the cutting depth before I begin planing. In most cases, a light, thin shaving is preferable to a thick, coarse one **(A)**. You'll discover that the plane is easier to push, and the surface will be smoother, with less tearout. The most precise method for adjusting the iron is to sight along the sole of the plane. Grasp the front knob with one hand while making adjustments to the depth screw and lateral adjustment lever with the other. When properly adjusted, the edge of the iron should just peek through the mouth of the plane. Use the lateral adjustment lever to correct any unevenness.

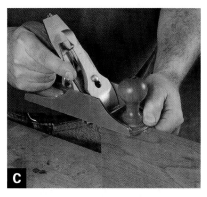
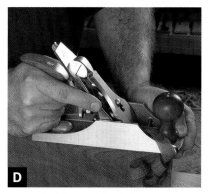

To start the cut, apply pressure to the toe of the plane **(B)**. Use your other hand to push the plane **(C)**. As you reach the end of the stock, push downward on the heel of the plane **(D)**.

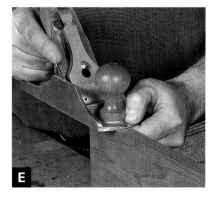
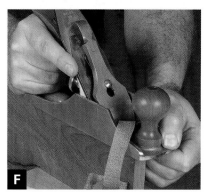

When pushing the plane, use your upper body as well as your arms. Begin each cut with your elbows bent **(E)**. As you push the plane forward, begin to extend your arms and lean forward into the cut. To complete the cut, keep your feet firmly positioned and extend your arms as you rock your body forward **(F)**. To keep the cut square, position your hand against the face of the board to steady the plane **(G)**.

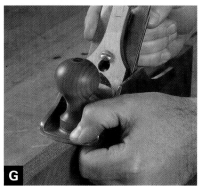

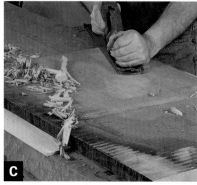

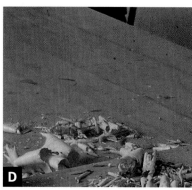

Flattening a Wide Board

I avoid gluing several narrow boards edge-to-edge to create a wide panel. The mismatched grain and color is just too distracting. Instead, I prefer to use one wide board for tabletops and door panels. It's best to flatten stock before planing it to thickness to remove any warp or twist. If your jointer is too small to accommodate anything beyond 6-in. or 8-in. wide, you can use a bench plane to flatten the board. A long plane, such as a No. 7 or No. 8, works best. The extra length will bridge the low areas, making it easier to flatten the board. And the extra weight of a long plane will help propel it through each cut with minimal chatter.

You'll want to use a heavy cut; remember that the purpose is to flatten the board, not to smooth it. I grind the iron convex and adjust the frog and iron for a coarse, heavy cut **(A)**. Afterwards, I use a No. 4½ bench plane to smooth the surface of the plank.

Begin by examining the board for twist with winding sticks **(B)**. A quick sight across the top of the sticks will reveal the high corners. Position stops against two adjacent edges and plane the board diagonally, from corner **(C)** to opposite corner **(D)**. After the initial planing, the highest points will be cut down.

Next, turn the plane slightly and broaden the areas cutting across the grain **(E)**. Continue planing until the two broad areas at each corner meet in the middle of the board **(F)**.

Finally, use a straightedge to find any remaining high spots and plane them away. When you're finished, the board should be flat and ready for planing to final thickness **(G)**.

Thicknessing a Board by Hand

To thickness a board with planes, begin by flattening one face of the board, The easiest and fastest method is to use a long bench plane, such as a No. 6, 7, or 8, and push the plane diagonally across the face of the board **(A)**. It's not necessary to smooth the board with the large, coarse cutting plane, only to flatten it **(B)**. Afterwards, smooth the flat face with a smoothing plane and carefully follow the grain **(C)**.

Next, set a marking gauge to the dimension of the final thickness and carefully scribe all four edges of the stock **(D)**. As you plane the second face, the scribe line will appear as a fine, feathery wisp of wood **(E)**.

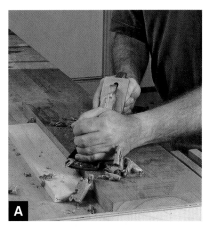

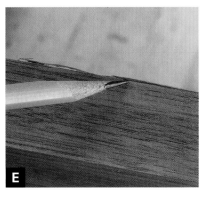

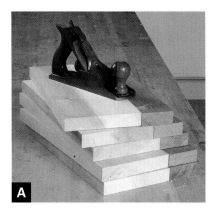

Planing Glued-Up Panels

After gluing panels edge-to-edge, I use a bench plane to flush the mating surfaces before planing the panel to thickness **(A)**. In this photo-essay, I'm planing panels created from walnut and poplar. The panels will be resawn to create dividers inside a spice cabinet. Because most of the divider will be hidden by drawers, I glued inexpensive poplar to the walnut primary wood to reduce costs.

If you're careful to match the grain direction in the mating boards, it will greatly simplify planing. Just plane in the direction of the grain **(B)**. If the grain changes direction in either board, it often works best to plane the stock diagonally **(C)** to minimize tearout **(D)**. Then finish smoothing the panel with a sharp scraper **(E)**.

[**TIP**] Dried glue can dull a plane blade, so always remove the glue squeeze-out before planing a glue-up panel.

Jointing an Edge

Edge jointing is the process of straightening the edge of stock prior to ripping or gluing it to another piece of stock. In the process of straightening the edge, it's also important that the edge is square to the board's face.

The sole of a plane is a great aid in truing the edge, provided that the sole is flat. However, the natural tendency is to hollow the edge of the stock when planing, or to remove more wood from the center of the cut. To compensate for this tendency, it's helpful to apply pressure at the toe of the plane when starting the cut **(A)**, then to transfer the pressure to the heel of the plane toward the end of the cut **(B)**. Use your thumb to apply downward pressure and your index finger to steady the plane against the stock **(C)**. If necessary, a final pass with an edge-trim plane will insure that the edge is absolutely square **(D)**.

Planing End Grain

End grain is often hidden by a strip of molding or a breadboard end. But when it's in full view, the surface will be dramatically improved by planing. End-grain fibers are tough and can often be difficult to work. The best plane to use is a very sharp, low-angle plane. A low-angle block plane will work, as will a longer low-angle bench plane like this one **(A)**. To prevent splintering of the trailing edge and help in keeping the surface flat and square, a shooting board is a great aid **(B)**.

➤ See *"Shooting Board"* on p. 52.

After cutting the stock to final length, position it against the stop of the shooting board and hold it firmly in place. With the other hand, push the plane with long, continuous strokes. If the iron is sharp and adjusted for a light cut, wispy see-through shavings will spill from the plane **(C)**, leaving the surface crisp and distinctive.

Planing Small Stock

During planing, the plane is usually pushed across the stock as it is held in place on the bench. However, for planing small stock, you may find it easier and more precise to reverse the procedure **(A)**. By placing the plane in the vise **(B)**, you can, in effect, create a very small jointer. You can even clamp a strip of wood to the sole of the inverted plane to serve as a fence. Begin by pressing the stock firmly to the toe of the plane **(C)**. As you push the work past the iron, keep pressure applied to the area nearest the iron **(D)**. Pushing the stock continuously from end to end will create a smooth, square, uniform edge **(E)**.

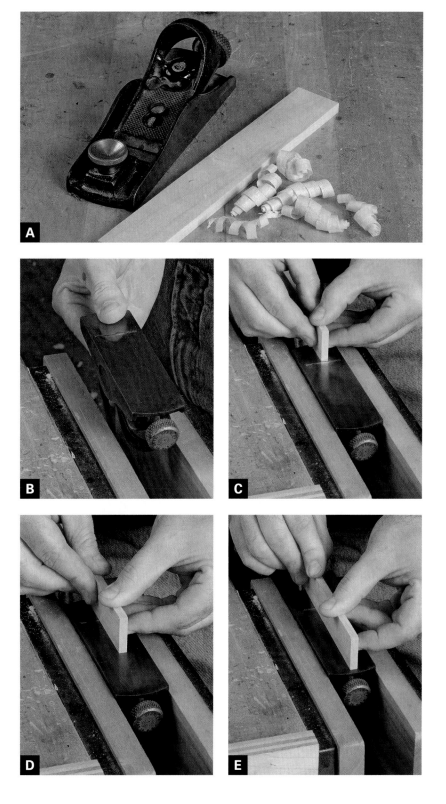

Smoothing a Tapered Leg

Machines and hand tools often make great companions; machines cut stock efficiently, and hand tools create the details. This tapered leg is a good example. After tapering the leg on the table saw, you can quickly smooth away the offensive saw marks with a bench plane **(A)**.

Begin by examining the leg to determine grain direction. In most cases, the grain will run toward the foot of the leg. Position the toe of the plane on the taper with the cutting edge slightly behind the taper **(B)**. Lean forward and push the plane with the force of your upper body **(C)**. As the plane approaches the end of the leg, apply downward pressure at the heel **(D)**. If necessary, skew the plane slightly to improve the surface quality **(E)**.

Fitting a Door

There are three types of doors: overlay, lipped, and flush fitting **(A)**. Flush-fitting doors are the most tedious to fit because, unlike the other two types, they must fit perfectly. Any variation between the door and the casework is easily spotted when the door is closed. A sharp, well-tuned bench plane makes achieving a clean, tight fit much easier.

When constructing the door, make the stiles and rails ½₂ in. wider than the final dimension. Also, make the door the same size as the opening. Position the door over the opening to check the fit. Most likely you'll need to work the stiles and rails to correspond to the cabinet **(B)**. Mark a stile and the adjacent rail where they overhang the cabinet. Secure the door in a vise and carefully plane a shaving or two from the marked area **(C)**. When planing the rail, work toward the middle of the door to avoid splintering the end grain on the stile **(D)**.

Once the trimmed stile and rail correspond to the opening, place the door back on the cabinet. Mark the cabinet where the opposite stile **(E)** and rail **(F)** overhang. Next, measure the overhang **(G)**, and plane equal amounts from the door.

A

B

C

D

E

F

G

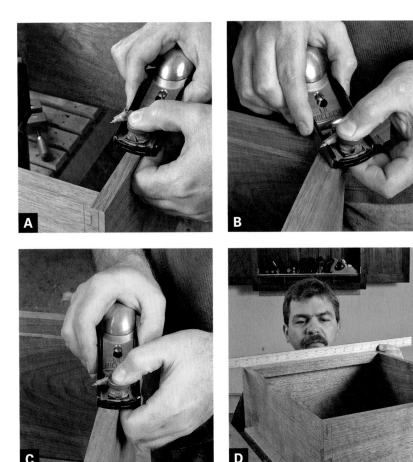

Planing Cabinet Edges

After assembly, furniture components must be smoothed and leveled to correct the inevitable surface variations that occur. This small spice cabinet is a good example. After the casework is dovetailed and assembled, one or two passes with a plane will level the intersections at the joints. Attempting to work these surfaces with sandpaper would leave them rounded and scratched at the intersections.

Begin by examining the case parts to determine grain direction. It's not unusual for grain to change directions somewhere along the length of the stock. As you plane, use your thumb to apply downward pressure to the toe of the plane and your index and middle finger to steady the plane and keep it running square **(A)**. As you approach the corner, skew the plane 45 degrees to avoid tearout at the intersection **(B)**. When you continue down the adjacent face, be prepared for the change in grain direction **(C)**. Afterwards, check the surfaces for true with a straightedge **(D)** and make any necessary adjustments.

Flush-Planing Dovetails

When cutting dovetails, I scribe the baseline larger than the thickness of the tail board. With this method, the pins are slightly proud after assembly. After I trim them flush, the joint appears crisp and clean, just the look I'm after.

Begin by securing the work to the bench. I used large handscrews to clamp this spice cabinet to the edge of my benchtop **(A)**. Now use a sharp, low-angle plane set for a light cut. As you shave the end grain of the pins, skew the plane so that the planing force is directed toward the cabinet **(B)**. This way you'll get a clean, crisp surface without splintering along the edges of the box **(C)**.

Planing a Rabbet

A rabbet is a recess along the edge of a board. It's a common woodworking feature that has wide applications, such as for the lipped edges of doors and drawers. Unlike shoulder planes, rabbet planes feature a fence and depth stop for controlling the dimensions of the final cut. To plane a rabbet, begin by setting the fence to the desired width **(A)**. Then adjust the sliding depth stop and lock it into place **(B)**.

Start the cut as you would with a bench plane, by applying pressure to the toe of the plane **(C)**. As you push the plane forward, extend your arms and lean into the cut **(D)**. As you approach the end of the cut, apply downward force to the heel of the plane **(E)**. Throughout the entire cut, use your index and middle fingers to apply pressure to the plane's fence; this will insure that the rabbet is full width **(F)**.

After repeated passes of the plane, the depth stop will rub the work and stop the cutting action. **(G)**.

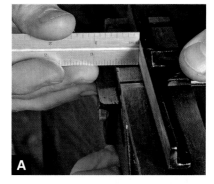

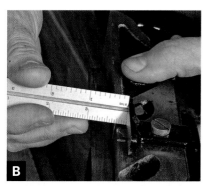

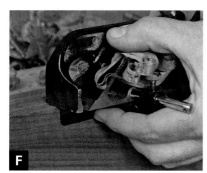

Planing a Chamfer

A chamfer is a decorative 45-degree bevel along the edge of the stock. Although you can cut a chamfer with the router, it's often more efficient to use a plane, especially if you have just one or two pieces to cut. Also, chamfering with a plane yields subtle inconsistencies that give your work a pleasant, handmade look.

Begin with layout. Use a pencil and a combination square to mark the limits of the chamfer along the edges **(A)** and face of the stock **(B)**. Chamfering is perfectly suited to a small block plane **(C)**. Starting with the ends, push the plane at a 45-degree angle with the plane skewed slightly downward **(D)**. As you approach the layout lines, adjust the angle of the plane if necessary.

After chamfering both ends, complete the job by planing the long-grain edges **(E)**. Saving the edges for last will insure that minor tearout at the corners will be planed away.

Planing a Quirk Bead

Wooden molding planes are still a great way to shape distinctive furniture moldings. This example is a quirk bead, commonly used along the edges of backs and face frames of cabinets.

For the best results, always select straight-grained stock and plane in the direction of the grain. As you plane, keep the fence firmly against the stock (**A**). Once the full depth of the profile is reached, the built-in stop on the plane will rub the stock (**B**).

Planing a Curve

Bandsaws excel at sawing curves, but like other machines, they leave behind a distinctive pattern that must be removed prior to finishing. Just as a bench plane is used to smooth away mill marks on flat stock, a compass plane can be used to smooth and refine curves (**A**).

Begin by adjusting the flexible sole of the plane to correspond with the curve to be planed (**B**). For this convex curve, I've adjusted the plane to be slightly flatter than the radius of the stock. As with any plane, follow the grain direction. When planing a curve, this means you'll want to start at the apex (**C**) and push the plane "downhill" (**D**).

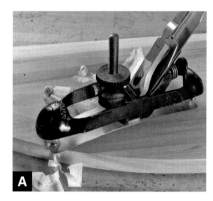

A

B

C

D

E

F

G

Tuning a Bench Plane

Attempting to use a poorly tuned plane can be a lesson in frustration as the plane chatters, or the iron digs in and the throat clogs with shavings. To get the best performance from any plane, you'll have to first tune it. The sole of a plane should be flat, but many are not. Unfortunately, many new planes have warped or improperly ground soles. And older planes may show signs of wear.

Flatten the sole by abrading it with a flat, coarse abrasive, such as a diamond lapping plate **(A)**. Stop periodically and check the sole for flatness with a machined straightedge **(B)**. Next, flatten the frog. Use a mill file to carefully smooth away any high spots or burrs that may be left over from casting **(C)**. The iron should rest firmly on the surface of the frog without rocking.

During planing, the iron also gets support from the cap iron, sometimes called a chipbreaker. The cap iron is a steel plate fastened to the back of the iron with a large screw. Consider replacing the iron **(D)** and cap iron with thicker components that have greater resistance to deflection **(E)**. The modest amount that you spend to replace the thin, vibration-prone iron and cap iron that came with your plane will dramatically improve the performance of the plane.

Next, check the fit of the cap iron to the iron. The cap iron plays an important supporting role for the iron. It reinforces the edge of the iron and curls the wood shaving as it exits the throat of the plane. To maintain consistent pressure near the edge of the iron, the cap iron should be slightly bent, or sprung, along its length and absolutely straight along its leading edge. If necessary, you can bend the cap iron by placing it in a vise and striking it gently with a mallet **(F)**. To straighten the edge of the cap iron, stroke it with a mill file **(G)**.

Be sure to retain a consistent bevel along the edge. To check the results, fasten the cap iron to the iron and torque the screw down firmly **(H)**. Now sight along the leading edge of the cap iron and look for traces of light, which indicates a gap in the fit **(I)**. For fine, wispy-thin shavings, the edge of the cap iron should be set approximately ⅟₃₂ in. below the edge of the iron **(J)**. Now sharpen the iron.

Finally, adjust the mouth of the plane by moving the frog. For smoothing, the mouth should be small—just enough for the finest shavings to pass through. Before reassembling the plane, flatten and polish the leading edge of the lever cap **(K)**. Then apply a coat of paste wax to all the parts and a drop of oil to moving parts, including the depth adjustment wheel and lateral adjustment lever. Adjust the iron for a light cut and an even shaving. Sight down the sole of the plane **(L)** and turn the adjustment wheel until the iron slightly protrudes **(M)**. To move the iron laterally, pivot the lateral adjustment lever located at the top of the frog.

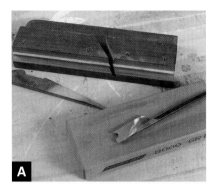
A

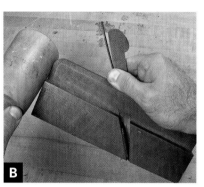
B

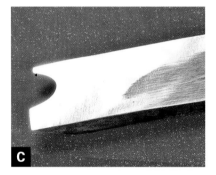
C

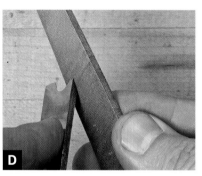
D

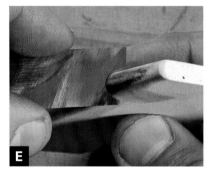
E

F

Tuning a Wooden Molding Plane

Old wooden molding planes are enjoyable to use, and they're still widely available. When shopping for old molding planes, avoid those with soles that display signs of excessive wear, or with pitted irons or cracked soles from exposure to moisture.

Begin by disassembling the plane. These unique tools are simple in design—just a body, wedge, and iron **(A)**. To loosen a stuck wedge, tap the heel of the plane gently with a mallet **(B)**. Begin by sharpening the iron. Polish the back to a mirror finish as you would any other plane iron **(C)**. The profile of the iron should match the wooden sole of the plane. However, repeated sharpenings by its former owner may have changed the profile. If this is not corrected, it may be impossible to set the iron to a consistent cutting depth. You can correct the iron profile by using small mill files **(D)**, followed by polishing with slipstones **(E)**. Fit the iron back into the plane and adjust it for a light shaving by sighting down the sole **(F)**.

Tuning a Spokeshave

To tune a spokeshave, first remove all the parts, including the thumbscrews used for adjusting the cutting depth **(A)**. Use a mill file to flatten the area of the casting that supports the iron **(B)**. Flatten and polish the leading edge of the lever cap **(C)** and sharpen the iron.

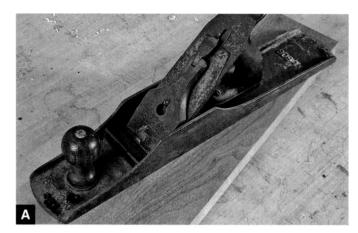

Restoring an Old Bench Plane

Here's a good example of what you can often find in the old tool market. This plane is a Stanley Bedrock No. 606 **(A)**. Although it's not a rare plane, any example of a Bedrock is uncommon, and the No. 606 is a desirable size. While the plane in these photos is somewhat rusty, it is intact, and the rust will not affect how it performs **(B)**.

Begin by disassembling the plane **(C)**. If the screws that secure the frog are frozen with rust, spray it with a lubricant **(D)**. You can get extra torque by placing a wrench on a square-shanked screwdriver **(E)**.

Place all the rusty parts of the plane in a pan of mineral spirits **(F)**. The mineral spirits help to loosen the rust as you clean the various parts with fine steel wool. A good sign—once all the dirt and rust was washed away, most of the Japanning (painted finish) was still intact. Best of all, the machined surfaces on the frog and sole still revealed their mill marks from machining years ago. The rust, although very distracting, was only on the surface.

After cleaning, reassemble the parts on the plane body. A drop of oil on the threads of screws and a film of oil on the machined surfaces of the frog make these parts work smoothly. The oil also helps prevent future rust **(G)**.

Although the iron in this plane was pitted with rust, it wasn't a concern. I had already planned to replace the thin Stanley iron with a cryogenically treated replacement iron. The new iron is also substantially thicker than the original, which, along with a new, improved cap iron, greatly adds to the performance of this fine old Bedrock plane **(H)**.

The frog position, sharpening, and tuning are all important elements of the performance of the plane.

▶ See *"Restoring an Old Bench Plane"* on the facing page, and *"Sharpening Hand Tools"* on p. 168.

After assembly, apply a coat of wax to the sole of the plane **(I)**. The wax will fill the microscopic pores in the cast iron to help prevent rust. The wax also lubricates the plane so it glides easily over the stock during use **(J)**.

Files and Rasps

Files and Rasps

➤ Smoothing a
Shaped Leg
(p. 167)

ILES AND RASPS CUT with rows of tiny teeth. Rasps are really coarse files with extra-large teeth for fast, aggressive shaping and sculpting of compound curves. Files are the smoother cutting cousins.

Sizes and Shapes

Open a catalog and you'll find a bewildering assortment of files from which to choose. File lengths typically range from the diminutive 4 in. to the huge 16 in. For smoothing surfaces in tight contours, you can purchase slender needle files.

Files are also categorized by the smoothness of the cut. The coarser cutting files,

bastard files (I'm not sure where the unusual name came from), are great for smoothing surfaces after shaping with a rasp. After smoothing with a bastard file, I usually skip the next two in the series, second-cut and smooth files, and switch to using a scraper. Second-cut and smooth files cut too slowly and clog easily.

Files are also categorized by shape. The most versatile shapes are flat, half-round, and round. Although there are other shapes available, such as triangles and squares, I don't find them useful for woodworking. Of the various shapes, I find the half-round most useful; the flat side works well for convex curves, while the half-round face of the

Rasps and files are available in a wide variety of shapes and sizes.

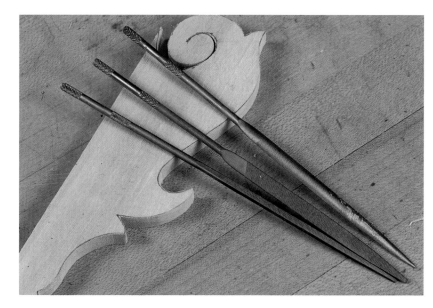

TWO TYPES OF FILES

Single cut—cuts slowly, leaves a smooth surface

Double cut—cuts more aggressively, leaves a coarser surface

For smoothing tight contours, select a slim needle file.

The fine-tooth pattern of a mill file works well for dressing the edge of a scraper.

file easily shapes and smoothes concave curves. By having both shapes on one file, you can quickly switch back and forth while smoothing.

Files are also categorized by the tooth design. Single-cut files cut more smoothly than the more aggressive double-cut type. Mill files have rows of fine teeth and work well for the initial dressing of scrapers to remove the old burr. Rifflers are special files

Rifflers are curved along their length.

The short, stiff wire bristles of a file brush are specially designed for cleaning files.

This file handle has threads that seat firmly on the tang of a file.

For safety and greater control, equip your files with handles.

with compound curves; they are ideal for smoothing tight areas in carvings. As files become clogged, you can clean them with a file brush. This special file brush has short, stiff wire bristles that are well-suited to this task.

When not using your files, store them separately in a tool roll or a wall-mounted rack to keep them sharp. Files tossed together in a drawer or toolbox will dull

quickly. Finally, equip all your files with handles. Equipped with handles, files are safer to use and the handles will also provide you with greater leverage. The best file handles have hardened steel threads that "bite" into the tang of the file and seat firmly.

Smoothing a Shaped Leg

Rasps and files are simple yet versatile tools that excel at sculpting and smoothing compound curves. These tools are not intended for creating curves; the curve should first be created by sawing, steam bending, or laminating.

When the curve is bandsawn, the rasp is often used first to remove minor imperfections. In order to smooth away bumps and dips, position the rasp askew to the workpiece **(A)** rather than at 90 degrees. During sculpting, the rasp is pushed in two directions simultaneously—downward with the grain along the workpiece, and forward, stroking the tool across the workpiece **(B)**. Use the flat side of the rasp for shaping convex surfaces and the convex face of the rasp for concave surfaces.

When shaping tight curves, roll the rasp with your wrist as you push it through the stock **(C)**. For speed and efficiency, it works best to turn the rasp end-for-end and pull it **(D)** rather than to continually reposition the work. With practice, you'll be able to blend and sculpt compound curves easily and efficiently **(E)**.

Once you're satisfied with the shape you've created with the rasp, switch to a file. The same techniques apply to both tools; the finer teeth **(F)** of the file will quickly smooth the coarse surface from the rasp.

Sharpening Hand Tools

Sharpening Chisels

➤ Sharpening a Chisel (p. 175)

Sharpening Scrapers

➤ Sharpening a Cabinet Scraper (p. 177)

S HARPENING MACHINE TOOLING, such as carbide-tipped sawblades and router bits, is best left to the experts. Professional sharpening shops have both the equipment and knowledge and will handle these maintenance chores for a modest fee. Sharpening hand tools is another story. Possessing the skills to hone chisels, plane irons, and carving tools to a razor edge is an essential part of woodworking. Sometimes a chisel or other edge tool is inadvertently nicked as it comes into contact with a square or other tool lying on the benchtop. Even under normal use, edge tools dull rather quickly and require continual mainte-

nance for the greatest control. Fortunately, it's not difficult to develop sharpening skills. Sharpening is simply a gradual process of removing steel with abrasives. As increasingly finer abrasives are used, the scratches in the steel become finer, the steel more polished, and the edge sharper.

There are many types of tools and abrasives from which to choose, from waterstones to sandpaper, and the type you select is largely a matter of personal preference. Although the various stones and grinders have different working characteristics and costs, all will adequately sharpen steel. Let's take a closer look at what's available.

There are a variety of abrasives, burnishers, and other tools available for sharpening a dull edge.

The larger wheel of this grinder (on right side) spins in a water bath that keeps the steel cool.

Grinders

Sometimes it is necessary to grind the steel to restore the edge of a chisel or plane iron. Perhaps the edge is nicked or worn from repeated honings. The most efficient way to restore the edge is with a grinder. You can choose between a common bench grinder or a more expensive wet grinder. Bench grinders require greater skill and a light touch to avoid overheating the steel and destroying the temper or hardness. Also, the tool rests on bench grinders are not well-designed for supporting woodworking tools. However, if you construct tool supports, up-grade your bench grinder with cooler-running aluminum-oxide wheels, and frequently cool the tool edge by dipping it in water, a bench grinder can be an effective, inexpensive machine for restoring the edges of your planes, chisels, and other edge tools.

Today there are a number of wet grinders available featuring large-diameter, wide grinding wheels that run in a water bath. Wet grinders have several advantages over standard bench grinders. The biggest advantage is that the continuous water bath over the wheel and tool edge makes it impossible to overheat the steel. Also, the wide wheel of a wet grinder is more effective for sharpening broad edges, such as the iron from a bench plane. Because wet grinders are designed specifically for woodworking tools, the tool rests usually provide better support than those typically found on bench grinders. The only real disadvantage to wet grinders is their higher price tag. Of course, grinding leaves a coarse surface, so it's vital to follow up by honing the tool edge with finer abrasives.

Bevel Angles

The first step in sharpening is to grind the bevel. I typically use a hollow bevel because it is faster and easier to hone than a flat bevel. Bevel angles usually range from 20 to 30 degrees, with 25 degrees being standard. A shallow bevel is sharper and creates less cutting resistance but may fracture more easily. A steeper bevel is more durable but has more resistance and provides less control. I have several sets of chisels in my kit, and I grind them according to how I plan to use them. For example, I grind a steeper angle on the short chisels I use with a mallet, and a shallow bevel on paring chisels.

Also, it's sometimes necessary to adjust the bevel angle slightly to compensate for the hardness of the steel. For example, if the edge fractures easily, try grinding a steeper bevel.

Edge Profiles

Plane irons are ground either square or convex, depending on the type of plane and the work that you're producing. Planes for joinery, such as rabbet and shoulder planes, should be ground square. The iron of a smooth plane works best when ground slightly convex. This profile prevents the corners of the plane iron from creating ridges in the board.

Bench planes used for flattening rough stock are called scrub planes. But any bench plane can be used as a scrub plane if you grind a pronounced convex profile on the iron. Just remember to adjust the frog of the plane for a coarse shaving.

GRINDING EDGE TOOLS

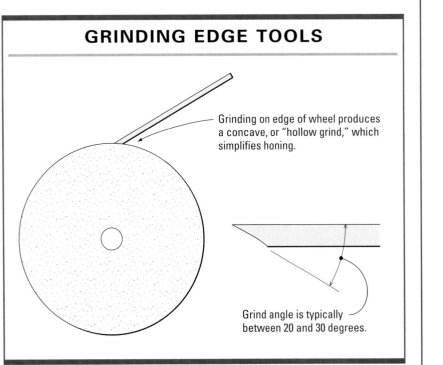

Grinding on edge of wheel produces a concave, or "hollow grind," which simplifies honing.

Grind angle is typically between 20 and 30 degrees.

BEVEL ANGLES

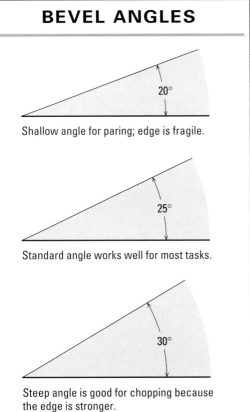

Shallow angle for paring; edge is fragile.

Standard angle works well for most tasks.

Steep angle is good for chopping because the edge is stronger.

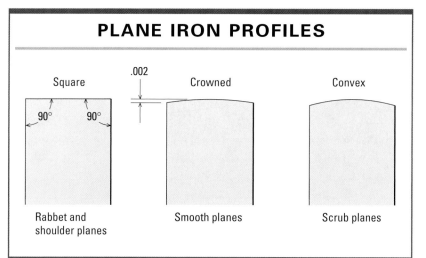

PLANE IRON PROFILES

Square — 90° 90° — Rabbet and shoulder planes

.002 — Crowned — Smooth planes

Convex — Scrub planes

Arkansas stones require oil as a lubricant.

Chisels are usually ground square to the edge, but skewed chisels are useful for many tasks, especially cutting dovetails. You can buy skewed chisels, but you can also regrind a skewed edge on a regular chisel.

Tools for Honing

After the grinding to restore the bevel or remove a nick in the edge, the next step is honing. Honing is the process of abrading the edge of the tool over progressively finer abrasives. The back of the tool should be polished, too; it converges with the bevel to create the cutting edge. However, the best time to polish the back of the tool is after you first purchase it. Later on, each time you hone the tool, you can concentrate your efforts on the bevel.

There are a number of tools available for honing. Man-made stones, such as synthetic waterstones, tend to cut quickly—but also wear quickly. In contrast, natural stones, such as Arkansas stones, wear slowly but become clogged easily and are tediously slow-cutting. Let's take a closer look at what's available so you can decide for yourself.

Arkansas Stones

Until recently, most craftsmen honed their tools with Arkansas stones. These natural stones are still quarried, graded, ground flat and square, and boxed ready for use. Natural Arkansas stones create a sharp, highly polished edge, and because they are hard, they retain their flatness for long periods of use. However, because Arkansas stones use oil as a lubricant, they tend to be messy. But their main drawback is that they cut so slowly. Sharpening a chisel to a razor edge with Arkansas stones is time-consuming in comparison to newer alternatives, such as manmade waterstones, so I retired my Arkansas stones years ago.

Ceramic Stones

I remember back several years ago when ceramic sharpening stones débuted. They were heralded as both hard and fast-cutting—great qualities for a sharpening stone. Additionally, because they were used dry, they were not as messy as other types of stones. When they became dirty with use,

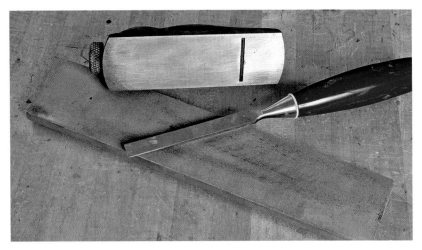

This "stone" is actually diamond particles bonded to a steel plate.

Waterstones cut very quickly compared to other sharpening stones.

A diamond plate works well for flattening a sharpening stone.

they could be easily cleaned with a mild household cleaner and water. However, I soon became disappointed with these stones; after several cycles of cleaning and use, their cutting ability gradually diminished.

Diamond Plates

Another form of sharpening "stone" is actually diamond particles bonded to a hard plastic block or flat-ground steel plate. Diamond cuts quickly, and because of its hardness, it can retain its cutting ability for years. A 600-grit diamond plate is great for flattening the back of edge tools and the soles of planes. However, most diamond sharpening stones are too coarse for the final stages of honing. A great option, though, is to use diamond paste on a hardwood block for the ultimate in sharpness.

Waterstones

These days, many woodworkers prefer waterstones for their fast-cutting features. First imported from Japan, waterstones are available in a wide variety of grit sizes, and the finest stones will produce an incredibly sharp edge. Because they use water as a lubricant, they are not as messy as stones that use oil. The only real downside to waterstones is that they wear quickly, so they must be reflattened often. Norton Abrasives offers its own line of waterstones, which also cut quickly but wear slowly in comparison to traditional waterstones.

Sandpaper

Still another choice for honing is sandpaper. You just need to back up the sandpaper with a truly flat surface, like plate glass. The wet or dry abrasive sheets sold at auto paint

Sandpaper backed up by plate glass has become a favorite sharpening system for many woodworkers.

For the final polish, a leather strop works well.

stores are your best choice; just moisten the paper and position it on the plate glass. As it wears, just toss it out and replace it with a fresh sheet. Best of all, you'll never need to flatten it. This is a simple solution to sharpening that has become quite popular with a number of woodworkers.

Strops

For the final polish of the edge, it's tough to beat a leather strop. Although you can purchase a strop from many woodworking tool catalogs, it's much less expensive to make your own. You can purchase a scrap of leather from your local shoe repair store and glue it to a block of hardwood. Charge the strop with rouge, a very fine polishing abrasive in a wax stick.

Slipstones

Slipstones feature thin, contoured surfaces for honing the curved edges of turning and carving tools. For the final polish after using slipstones, flex a small piece of leather to conform to the shape of the gouge.

Slipstones come in many shapes for honing gouges.

To polish the edge of a gouge, you can flex a strip of leather to conform to the contour.

SCRAPER HOOKS

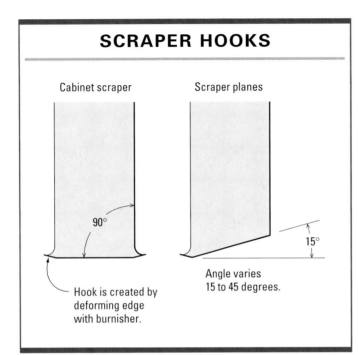

Cabinet scraper

Scraper planes

90°

15°

Angle varies
15 to 45 degrees.

Hook is created by
deforming edge
with burnisher.

A magnifying glass or photographer's
loupe will allow you to spot defects in the
cutting edge.

Cabinet Scrapers

The cabinet scraper is a great tool for
smoothing wood and reducing sanding to a
minimum. If you find it difficult to sharpen,
you're not alone. Scrapers cut with a tiny
burr that is formed after honing. The key to
sharpening a scraper is to avoid rounding the
edge while honing it. It's impossible to form
a hook on a rounded edge.

Scrapers are first sharpened, then bur-
nished to create a burr. The angle of the
scraper edge depends upon the type of scraper
and how it will be used. The edge of a cabinet
scraper is filed 90 degrees to the face. For
scraper planes, an angle of 15 to 20 degrees
works well. After you polish the face and
edge, a burnisher is used to draw the burr.

Unlike other edge tools, scrapers cut with a
tiny hook or burr. So after polishing a scraper,
you'll need a burnisher to hook the edge.
The traditional burnisher is simply a hard,
polished-steel rod equipped with a handle.
Newer-style burnishers feature a small carbide
rod mounted in a wooden or plastic block.
The block serves as a jig to maintain the correct angle as the burr is created.

▶ SHARPENING HANDSAWS

If you protect the teeth of your handsaws as you use them
and store them in a safe place, you'll find that they don't need
to be resharpened often. In fact, it may be several years
between sharpenings. Sharpening handsaws isn't difficult, but
it is tedious and time-consuming. If there's a reputable saw-
sharpening shop in your area, you may decide to let them
handle this job. Japanese saws have a complex tooth geometry that is best left to someone with experience. Fortunately,
many Japanese saws have disposable blades that fit into the
frame of the saw. Coping-saw blades are also disposable and
quite inexpensive.

Sharpening a Chisel

Sharpening a bench chisel is a good way to begin developing your skills. Remember that it always requires two surfaces to create an edge. When sharpening a chisel, you must polish both the beveled edge and the back. Also, the back of a chisel should be absolutely flat. The back often guides or registers the cut; if the back is rounded, the accuracy of the cut will be lost.

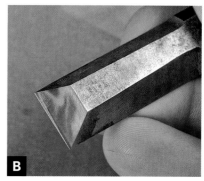
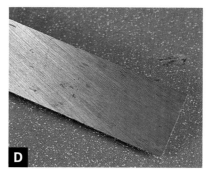

Sharpening is a two-step process. First grind the bevel, then hone the edge. Remember, grinding quickly removes a lot of steel. This is the best option when the edge has been damaged **(A)** or the hollow bevel has been worn away through repeated honing **(B)**. However, it's not always necessary to grind an edge. In fact, in most cases, an edge can be honed back to sharpness in just a few minutes with bench stones.

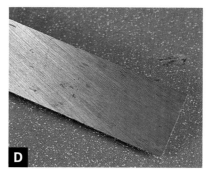

As you grind the edge, keep it cool to avoid annealing or softening the steel. If you're using a dry grinder, dip the tool in water often. Also, check the bevel angle as you grind and adjust the tool rest if necessary **(C)**.

Next, examine the back of the cutting edge. If the tool is a recent purchase, it may be necessary to remove grind marks, or, if it's an older tool, it may be marked by surface rust **(D)**. Starting with a coarse stone, flatten and polish the back to a mirror finish. You only need to polish about an inch or so back from the cutting edge **(E)**.

The next step is to hone the bevel. Be careful not to become overzealous and hone too much. An edge can usually be honed several times before it needs to be reground. However, honing too much steel away initially shortens the life of the edge.

To begin honing, position the bevel on a fine stone so that the edge and heel are both in contact **(F)**. You may find it helpful to rock the bevel until you feel the edge and heel make contact.

(Text continues on p. 176.)

Lock your wrist and glide the bevel across the surface of the stone with long, smooth strokes (**G**) until you can feel a slight burr across the back of the edge. You can also buy a simple guide wheel that holds the blade at a consistent angle for honing. Then stroke the back of the chisel across your finest stone to remove the burr. The bevel should reveal a narrow shiny surface across the edge and heel (**H**). Finally, polish the edge further by passing it several times over a leather strop (**I**).

[**TIP**] **While grinding or honing, you can get a really close look at the edge with a slide viewer or jeweler's loupe that will magnify the cutting edge so you can easily spot the slightest defect.**

You can check the edge for sharpness by slicing a thin shaving from a soft wood, such as pine or poplar (**J**). When sharp, the tool will sever the end grain cleanly and reveal the cell structure of the wood (**K**). In contrast, a dull edge will crush the soft fibers.

Sharpening a Cabinet Scraper

A sharp scraper will smooth the most difficult grain without producing tearout. The secret is in the burr. Before creating the burr, the edge must be sharp and highly polished. Because cabinet scrapers are made from relatively thin steel, it is easy to inadvertently round the edge during honing, making it impossible to form the burr.

Begin by filing the edge with a smooth mill file to remove the old burr and square the edge. To file the edge, use a technique called draw filing; hold the file perpendicular to both the face and the edge of the scraper and "draw" or pull it toward you **(A)**. Surprisingly, the scraper will cut quite well at this stage, although it leaves the surface of the wood somewhat rough. For a cleaner-cutting scraper, you'll need to take the sharpening process further.

Next, hone the face and edge of the scraper with bench stones **(B)**. To keep the edge square, flex the blade with your thumbs as you push it back and forth over the bench stones **(C)**. Flexing the scraper effectively creates a broad footprint to prevent the scraper from tipping and rounding the edge.

The next step is burnishing. But first wipe away the stone residue from the surface of the scraper and apply a drop of oil to the edge. The oil will lubricate the burnisher to create a smoother edge. Hold the burnisher at an angle of 5 to 10 degrees, and with moderate downward pressure, push it across the edge several times until you can feel a small burr **(D)**. A burnishing jig will make the job easier and maintain a consistent angle **(E)**.

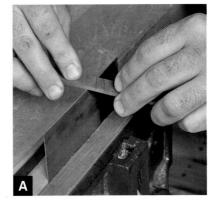

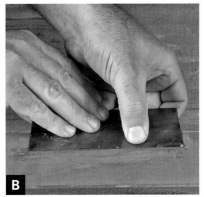

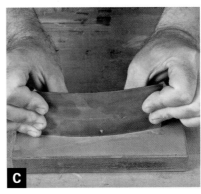

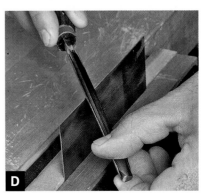

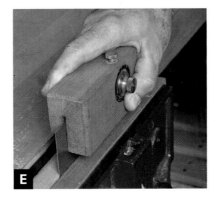

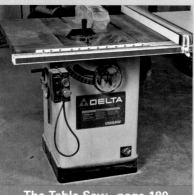
The Table Saw, page 180

The Jointer and Planer, page 201

The Bandsaw, page 215

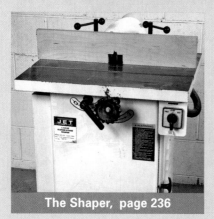
The Shaper, page 236

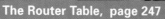
The Router Table, page 247

Drilling and Mortising, page 259

PART FOUR

Power Tools

I F YOU'VE EVER TRIED SMOOTHING a rough board entirely with a hand plane, you're sure to appreciate the efficiency of power tools. The jointer and planer remove the drudgery of smoothing a stack of rough boards by hand. The table saw rips and crosscuts stock to size, while the bandsaw cuts curves and resaws thick boards into perfectly matched thinner ones. And both saws can be used for cutting fine joinery. Add a router table (or shaper) and a drill press, and you've got a pretty complete power-tool woodworking shop. But power tools don't replace hand tools; they complement them. As an added benefit, power tools provide more time to enjoy doing fine, detailed handwork, such as carving a shell or cutting a dovetail joint. So, read on to find out what power tools are available, and how you can put them to work in your own shop.

The Table Saw

Tune-Up Procedures

➤ Checking Blade Alignment (p. 190)

➤ Aligning the Fence (p. 190)

➤ Changing Blades (p. 191)

Basic Operations

➤ Ripping to Width (p. 192)

➤ Ripping Narrow Stock (p. 192)

➤ Crosscutting on the Table Saw (p. 193)

➤ Crosscutting Short Stock (p. 194)

➤ Stop Cuts (p. 195)

Table-Saw Joinery

➤ Cutting Tenons on the Table Saw (p. 196)

➤ Dovetails on the Table Saw (p. 197)

Table-Saw Shaping

➤ Tapering with the Table Saw (p. 198)

➤ Cutting a Cove on the Table Saw (p. 199)

THE TABLE SAW IS A UNIVERSAL TOOL that forms the cornerstone of most woodworking shops. It efficiently rips and crosscuts stock to dimension, cuts joints, such as dovetails and tenons, and can even be used for shaping coves and raised panels.

Table saws cut with a circular blade mounted on a shaft or arbor, mounted under the table. With the handwheels on the sides of the cabinet, the blade can be raised or lowered to adjust for stock thickness and even tilted up to 45 degrees for angled cuts.

To guide the stock in a straight path past the blade, table saws use a fence. The fence rides a pair of rails and locks into position parallel to the blade for ripping. For cutting stock across the grain, a miter gauge is used. The miter gauge is guided in a straight path by a steel bar that slides in a parallel groove milled in the tabletop. The miter gauge works well for most cuts, but it is limited by its small size when you're crosscutting wide panels, such as the sides of casework. A great way to solve the problem is to construct a crosscut sled.

➤ See *"Building a Crosscut Sled"* on p. 184.

Although some small benchtop table saws are direct drive, most use pulleys and belts to transfer power from the motor to the blade. Contractor-type table saws use a single pulley system, while the more powerful cabinet

THE TABLE SAW

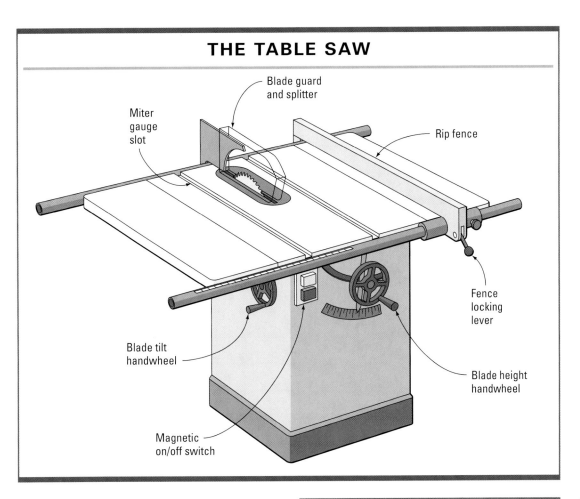

- Blade guard and splitter
- Miter gauge slot
- Rip fence
- Fence locking lever
- Blade tilt handwheel
- Blade height handwheel
- Magnetic on/off switch

saws feature larger motors equipped with a triple V-belt and pulley set-up.

Table Saw Guard and Splitter

The most important safety features of any table saw are the guard and splitter. Using any woodworking machine is inherently dangerous, but unlike other woodworking machines, such as the bandsaw, the table saw is prone to kickbacks. This phenomenon can occur whenever the stock contacts the back of the blade. The result is that the workpiece is thrown back violently toward the user. Obviously, if a hand is near the blade during a kickback, a serious injury can occur.

A splitter reduces the risk of kickback.

A guard is an important safety feature for any table saw.

This splitter conveniently snaps into position.

THE KICKBACK ZONE

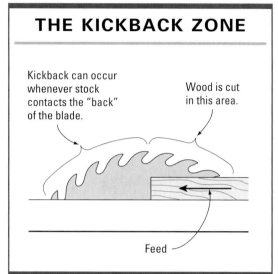

Kickback can occur whenever stock contacts the "back" of the blade.

Wood is cut in this area.

Feed

Fortunately, a guard and splitter will substantially reduce the risk of injury from using a table saw. The guard provides a barrier to shield your hands from the blade, while the splitter makes it virtually impossible for the workpiece to come into contact with the back of the blade.

Most new saws come equipped with a one-piece guard and splitter. This design is somewhat limiting, because the guard and splitter cannot be used separately. For example, when you're cutting a groove, the cut doesn't extend through the workpiece, so the splitter can't be used. However, with a separate guard and splitter, the guard can be used independently of the splitter and vice versa. This type of guard provides greater flexibility and an extra margin of safety. Although a two-piece guard and splitter is somewhat expensive, it's worth the extra cost for the additional safety it offers.

AVOIDING KICKBACK

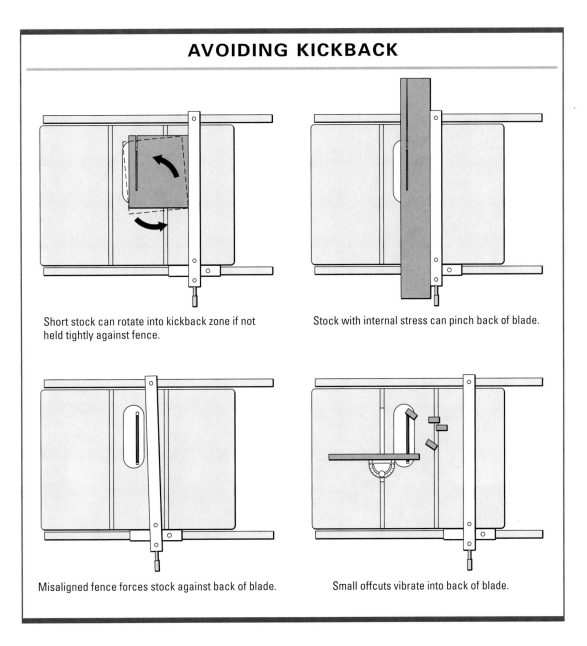

Short stock can rotate into kickback zone if not held tightly against fence.

Stock with internal stress can pinch back of blade.

Misaligned fence forces stock against back of blade.

Small offcuts vibrate into back of blade.

Within easy reach of your table saw should be a push stick. This simple device positions your hand a safe distance from the blade when you're ripping. You can purchase a push stick or make one of your own. Push blocks are also useful for cutting coves and other unusual setups at the table saw.

Ripping stock that has internal stress can also cause a kickback; as the stock is ripped, the stress is released and the workpiece may pinch and bind against the blade. A splitter will help with this problem, but the best solution is to avoid ripping stock that may be stressed. Areas around knots cause the blade to pinch and bind. Instead, use a bandsaw to rip problem lumber. The force of a bandsaw blade, in contrast, pushes the workpiece down against the table, and so it

► BUILDING A CROSSCUT SLED

Using the miter gauge to crosscut wide panels is difficult at best. The head of the miter gauge is too small to support the panel, and the bar that fits into the groove is too short. An easy, inexpensive solution is to construct a crosscut sled. The sled has two runners and a broad surface to support the panel.

Begin by cutting a sheet of ¾-in.-thick plywood to the size of your saw's table to serve as the base. Next, mill a pair of guide strips that fit snugly within the miter-gauge slots of your saw. To attach the guide strips, first lay them in the miter slots, position the plywood base on top, and drive a couple of small nails through the plywood into the guide strips. To ensure that the sled will cut square, align the edge of the plywood with the table edge. Now invert the sled, and permanently fasten the strips with countersunk screws.

Next, attach heavy strips at the front and back edges of the base. The back strip will serve to support the work during crosscutting, and both strips will serve to stiffen and strengthen the sled. Mill the strips large enough so that the sawblade doesn't cut through them when crosscutting 1-in.-thick panels. For safety, attach a plastic guard to the top of the sled and a thick block where the blade penetrates the back. Finally, test the sled to ensure that it is cutting 90 degrees, and, if necessary, loosen the screws that secure the back strip and make an adjustment.

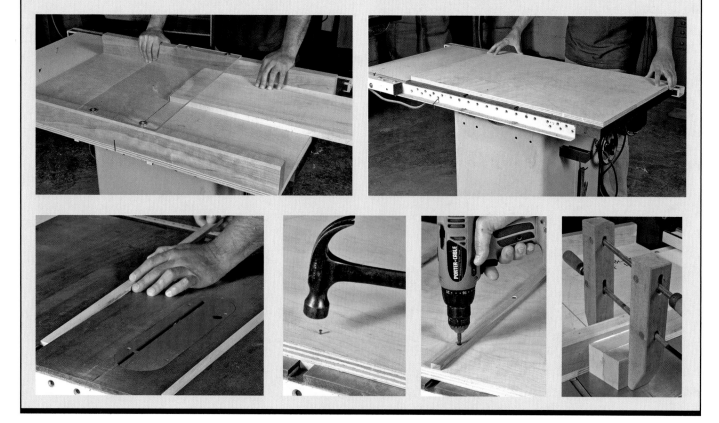

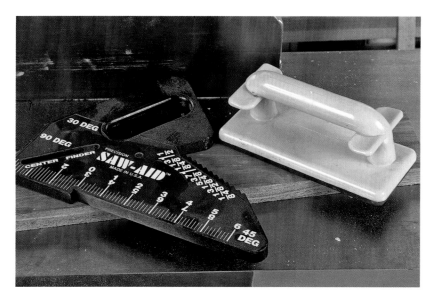

Push sticks and push blocks help position your hands a safe distance from the cutter.

can't kick back, even if the blade binds in the stock.

A misaligned fence on the table saw can also be a cause of kickback. When properly aligned, the fence should be parallel to the blade. The problem occurs if the trailing end of the fence is closer to the blade than the leading end.

When crosscutting, don't allow small off-cuts to accumulate near the back of the blade. And don't give in to the temptation to push them away from the blade with a pencil or stick. Instead, turn off the saw once a few cuts have been made, and clear the table before continuing.

Also, avoid ripping short stock. Any piece less than 12 in. in length can easily bind and kick back. And never rip stock that is wider than its length. There is simply not enough bearing surface against the fence to guide the stock past the blade. Instead, make it a habit to rip long stock and crosscut it to length afterwards. Finally, always wear eye and ear protection when using any power tools.

Always use personal safety equipment when working with any power tool.

▶ TABLE-SAW SAFETY GUIDELINES

- Always read and follow manufacturer's instructions.
- Never saw freehand. Always use the fence, miter gauge, or a jig that rides the fence or miter-gauge slot.
- Use a splitter whenever possible.
- Always use a blade guard.
- Use push sticks for ripping narrow stock.

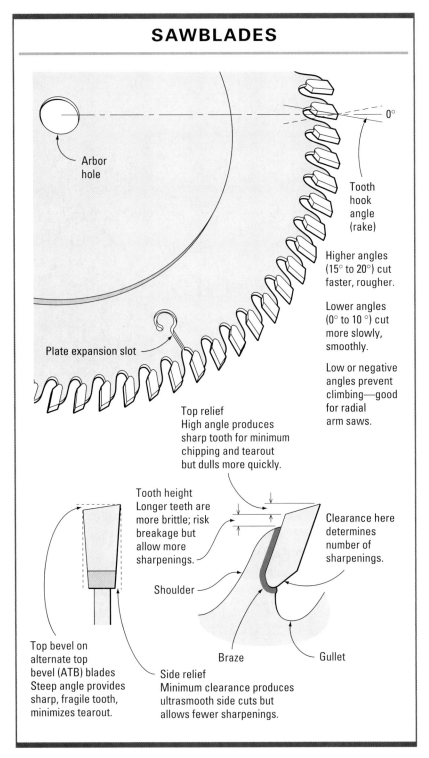

SAWBLADES

Arbor hole

Plate expansion slot

Tooth hook angle (rake)

0°

Higher angles (15° to 20°) cut faster, rougher.

Lower angles (0° to 10 °) cut more slowly, smoothly.

Low or negative angles prevent climbing—good for radial arm saws.

Top relief
High angle produces sharp tooth for minimum chipping and tearout but dulls more quickly.

Tooth height
Longer teeth are more brittle; risk breakage but allow more sharpenings.

Clearance here determines number of sharpenings.

Shoulder

Braze

Gullet

Top bevel on alternate top bevel (ATB) blades
Steep angle provides sharp, fragile tooth, minimizes tearout.

Side relief
Minimum clearance produces ultrasmooth side cuts but allows fewer sharpenings.

Sawblades

The heart of any saw is the blade. A great blade can coax more precision and performance from an average saw, but a great saw will disappoint you when equipped with a low-quality blade. These days, there are a vast number of blade types, tooth grinds, and hook angles, but remember, too, that there are trade-offs with any blade. For example, a greater number of teeth will produce a smoother cut, but the teeth will also require more horsepower and have a tendency to burn. Besides, accuracy is generally more important than smoothness; regardless of how smoothly a machine cuts, it's important to remove machine marks by handplaning, scraping, or sanding, or a combination of all three. Let's examine the designs and wade through the terminology so that you can make the best decision when purchasing sawblades.

These days, carbide has largely replaced steel as the material for sawteeth. The reason is clear—carbide is much harder and retains its edge up to 20 times longer. During manufacturing, carbide tips are brazed onto a steel body and then ground for both performance

For the best performance, equip your machines with a quality sawblade.

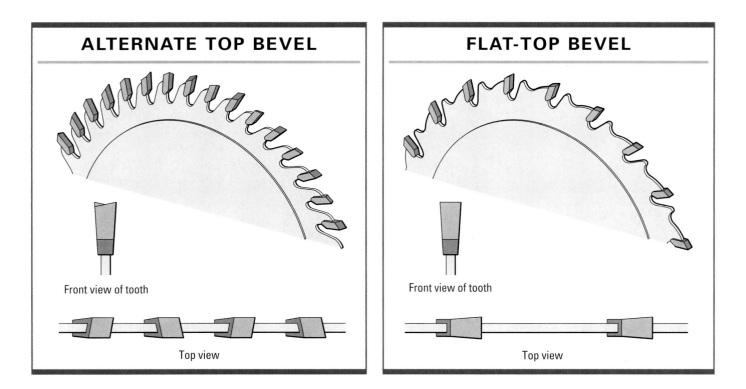

ALTERNATE TOP BEVEL

Front view of tooth

Top view

FLAT-TOP BEVEL

Front view of tooth

Top view

and sharpness. The number of teeth and the angle of grind on the face and top of each tooth have a profound affect on how the blade performs. As you might expect, sawblade manufacturers continually experiment with different grinds and combinations of grinds. The most common are flat-top bevel (FTB), alternate-top bevel (ATB), and alternate-top bevel with raker (ATB&R). These terms all refer to the grind, or bevel angle, at the top of each tooth. The angle that is formed by the intersection of the face of each tooth and an imaginary line from the blade's center is referred to as the hook angle. Hook angles range from an aggressive 20-degree angle found on many rip blades to an angle of minus 5 degrees, which limits climbing associated with miter saws and radial-arm saws.

A flat-top grind is essentially a square tooth. For fast, efficient ripping, manufac-turers use an aggressive hook angle, usually 20 degrees, a flat-top grind, and few teeth, often around 24, on a 10-in. diameter blade. Ripping consumes horsepower, and so a blade with fewer teeth will require less horsepower than one with more teeth. In between the teeth are large spaces, called gullets, which efficiently transport the saw-dust from the kerf created by the teeth.

Clean crosscutting requires a dramatically different design. The more teeth that are in contact with the wood, the smoother the cut. So crosscut blades will have as many as 80 teeth. To shear the tough end-grain fibers, the tops of the teeth are ground with a bevel to create knifelike edges that score the wood. Alternate teeth are ground in the opposite direction and the face is typically ground at a smooth-cutting 5 degrees.

Combination blades are a compromise between the fast-cutting rip blades and smooth crosscutting blades. Like crosscutting blades, combo blades employ an ATB grind to shear the stock cleanly during crosscutting, along with an occasional flat-topped raker tooth to help clear the way during ripping. Most combination blades are 50-tooth, again a compromise, and the teeth are arranged in 10 groups of 5—4 ATB teeth and 1 raker. In between the groups are large gullets that aid in clearing the sawdust during ripping.

Combination blades are my favorites for the table saw; remember, the table saw is a universal machine. Combo blades perform both ripping and crosscutting quite well and will allow you to use the saw without consuming time with frequent blade changes. You may find that when ripping large quantities of dense hardwoods, oak or maple, for example, the table saw may have a tendency to overheat. Switching to a rip blade will solve the problem. Some manufacturers tout the smoothness of their sawblades, and the price tag reflects that. However, when you're building furniture and other fine woodwork, it's important to remove all traces of saw marks by handplaning, scraping, or sanding, or a combination of the three. So for most woodworkers, the extreme smoothness of the most expensive blades is simply not worth the extra cost. However, I also avoid the low-end, do-it-yourself blades found in many home centers. The lower price is a result of inferior carbide and stamped bodies that are often warped. These blades are no bargain.

Thin-Kerf Blades

Thin-kerf blades create a kerf that is marginally less than that of a standard blade, not really a significant amount of lumber saved. However, thin-kerf blades require less horsepower than standard-thickness blades. So when ripping a large stack of hard maple, you're sure to notice the seeming increase in your saw's power. Be aware, however, that the narrow kerf may bind on your table saw's splitter.

For accurate crosscuts, I stick to a standard-thickness blade. The body of a thin-kerf blade tends to deflect in the cut and spoil the accuracy.

Dado Heads

Another type of sawblade is the dado head. This saw cuts a wide kerf for making joints, such as grooves, dadoes, and tenons. There are two types of dado heads, wobble and stacking. A wobble head adjusts for the width

A stacking dado head is invaluable for making a number of joints on your table saw.

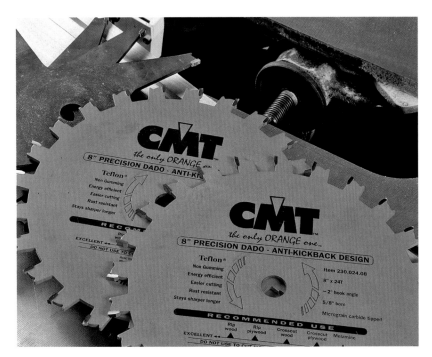

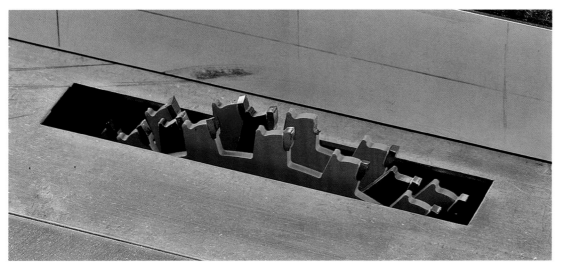

You'll need a special throat plate with a wider opening to use a dado head on your table saw.

of the groove by rotating a wedge-shaped washer on the side of the head. Because of this unusual design, this inexpensive tool cuts a groove with a rounded bottom, so it's definitely not the choice for fine work.

The best dado design is a stacking set. The tool comes as a set of two outer blades and chippers, a stack of cutters that mount on the arbor between the outer blades. The width of the groove can be adjusted from $1/4$ in. up to $13/16$ in. by adding more chippers.

As you mount the head to the saw's arbor, it's crucial to stagger the chippers so that the carbide tips don't touch. Also, whenever the dado head requires sharpening, take the complete set to the sharpening shop. This way, as the carbide is ground, all the chippers and the two outer saws will remain the same diameter.

To use a dado head, you'll need to purchase a special throat plate from the manufacturer of your table saw. It will have a wide opening to accommodate the head. Also, the wide basket found on most aftermarket table-saw guards can be used to provide an added measure of safety. Unfortunately, the guards that come equipped as part of many new table saws are not wide enough to fit over a dado head.

Table Saw Tune-Up

You'll get smoother, safer cuts from your table saw if you take a few minutes to tune it up. For the best performance, the blade and the fence should be parallel to the miter-gauge slot. Check the blade first; afterwards, check the fence alignment.

It's also a good idea to check belt tension periodically; as belts wear, they have a tendency to stretch, and you'll notice a loss of power. Finally, check the 90-degree and 45-degree blade stops. They each comprise a bolt with a jam nut to lock it in place. Although the stop may have vibrated out of adjustment, it may just need cleaning; it's usually only compacted sawdust that is creating the problem.

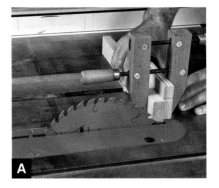

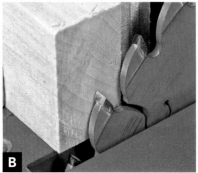

Checking Blade Alignment

For smooth, safe ripping, the blade and fence on your table saw must be parallel. Start with the blade. To check alignment, first clamp a freshly crosscut stick to the miter gauge. Before tightening the clamp, carefully position the end of the stick against a sawtooth **(A)**. For greatest accuracy, select a tooth that is pointing toward the stick **(B)**. Next, slide the miter gauge and stick toward the back of the blade and check a second tooth pointed the same way, toward the stick **(C)**. The difference in measurement, if any, can be corrected by loosening the bolts that secure the top to the saw's base and repositioning the top. Also, for this alignment technique to work effectively, the miter gauge must fit firmly in its slot, and the blade must be flat.

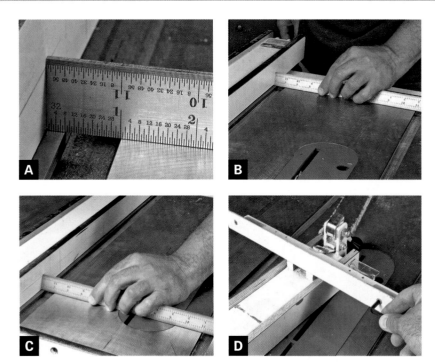

Aligning the Fence

After ensuring that the blade is parallel to the miter-gauge slots (see photo-essay above), lock the fence into position a short distance from one of the slots. Measure the distance with an accurate, engraved rule **(A)**. A stick rule or tape measure, both of which have painted lines, lack the necessary accuracy for this measurement. Check the distance at the front end **(B)** and the back end of the fence **(C)**. If necessary, make an adjustment to the fence locking mechanism to correct any difference in the two measurements **(D)**.

Changing Blades

To change the blade on your table saw, first disconnect the machine from its power source. Because table saws typically don't have a spindle lock, you'll need to brace the blade when removing the spindle nut. A block of wood works well and doesn't risk damaging the brittle carbide teeth **(A)**. Grasp the stick with one hand while pulling the spindle wrench toward you **(B)**.

Install the blade with the teeth facing toward you, and slide the washer into place **(C)**. Finger-tighten the spindle nut first; then position the spindle wrench against the table, carefully grasp the blade, and pull it toward you firmly to snug it down **(D)**.

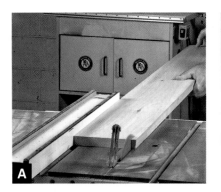

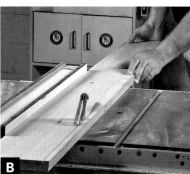

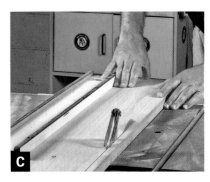

Ripping to Width

Ripping is the process of cutting a board along the grain to reduce its width. It's a common technique that's used for sizing stock prior to cutting joints and shaping.

Before beginning, joint the edge of the stock with a jointer or bench plane. After setting the fence for the desired width, position the jointed edge of the stock against the fence. Push the stock with one hand while using your other hand to maintain pressure against the fence **(A)**. As you push the stock forward, pause for a brief moment to change hand position **(B)**. As you approach the end of the cut, make certain to keep both hands out of the path of the blade **(C)**. For cuts of less than 6 in., use a push stick to finish pushing the stock through and past the blade. (See next photo-essay.)

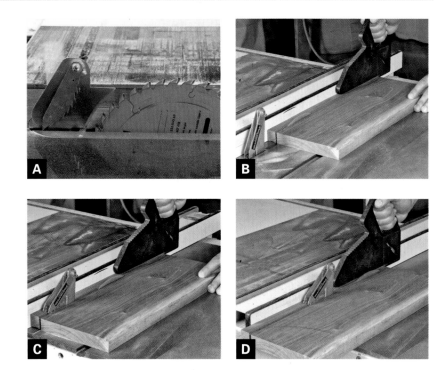

Ripping Narrow Stock

Ripping narrow stock on the table saw presents two challenges—maintaining accurate and consistent stock width, and avoiding kickback. Remember, narrow stock is kicked back with much greater velocity than large panels. To rip narrow stock safely, always remember to use a splitter **(A)** and a push stick. As you feed the stock, maintain pressure against the fence in front of the blade **(B)**, not adjacent to the blade, where pressure will cause the wood to bind. Also, apply feed pressure with the push stick against the portion of the workpiece that is riding the fence **(C)**. As the end of the stock approaches the blade, continue pushing only with the right hand **(D)**. Continue pushing the workpiece well beyond the blade and splitter.

Crosscutting on the Table Saw

Crosscutting on the table saw is usually performed with a miter gauge. Before cutting the workpiece, it's a good idea to cut a test piece and check the miter gauge for squareness.

For greater support of the workpiece, I attach a strip of wood to the face of the miter gauge (**A**). Next, saw a kerf into the backing board (**B**). Align the layout mark with the kerf (**C**). Now grasp the workpiece with one hand and the gauge with the other and make the cut (**D**). When cutting multiple pieces to the same length, clamp a stop to the opposite end of the backing board to insure accuracy (**E**).

> ⚠ **WARNING** Don't let small off-cuts gather at the back of the blade. Turn the saw off periodically and remove the off-cuts.

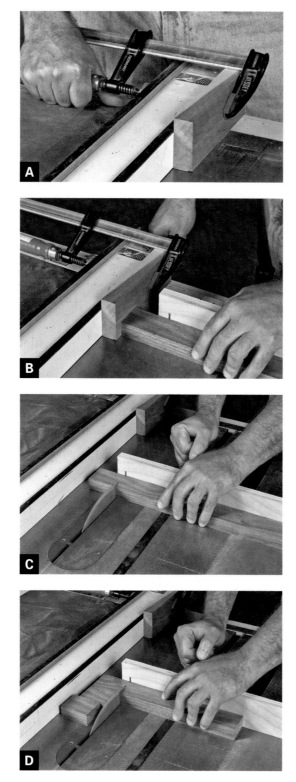

Crosscutting Short Stock

For obvious safety reasons, always cut short stock from a longer workpiece to distance your hands from the sawblade. And never use the fence as a stop; the work will bind between the blade and the fence and kick back. Instead, clamp a thick block to the fence a few inches behind the front of the blade to serve as a stop **(A)**. To make the cut, first position the workpiece against the stop block **(B)**, then make the cut **(C)**. As the pieces are cut, stop the saw occasionally and remove them from the table **(D)**.

Stop Cuts

A stop cut is a ripping operation that terminates before reaching the ends of the stock. A common use is for making the cutout in a cabinet base **(A)**. First, lay out the cut lines on the workpiece. Now position the stock adjacent to the blade and raise the blade so that the teeth just clear the stock **(B)**. Next, position the fence next to the blade, mark the points where the blade enters and exits the table **(C)**, and lock the fence into position for the width of the stop cut **(D)**. Before making the cut, you'll need to lower the blade completely. Align the layout mark on the workpiece with the point on the trailing end of the fence **(E)**, and clamp a stop block to the fence **(F)**.

> ⚠️ **WARNING** Never attempt a stop cut by lowering the workpiece onto the blade. Always raise the blade up through the workpiece, and use a stop block to prevent kickback.

Now you're ready to make the cut. Position the workpiece against the stop block, hold the workpiece firmly to the table, and raise the spinning blade until it clears the stock **(G)**. Now feed the stock until the layout line on the stock corresponds with the second line on the fence **(H)**. Keep in mind that the blade cuts further on the bottom face of the stock than on the top, where your layout lines are. Finally, make the crosscuts at each end to complete the process **(I)**.

A

B

C

D

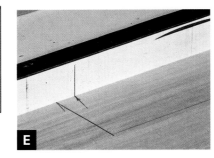

E

F

G

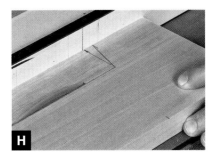

H

I

Cutting Tenons on the Table Saw

Mortise-and-tenon joinery is the strongest method for making door frames, casework, and even chairs. The faces of the tenons provide plenty of glue surface area, and the tenon shoulders resist racking (**A**).

Begin by disconnecting the table saw from its power source and mounting a dado head (**B**). Next, set the height of the dado head to remove the stock on one face of the tenon (**C**). For the greatest accuracy, position the fence for use as a stop. Simply align the layout mark for the tenon shoulder with the outermost portion of the dado head and lock the fence into place (**D**).

Starting at the end of the stock (**E**), make a series of overlapping crosscuts (**F**), using the miter gauge to guide the stock. Position the end of the stock against the fence to accurately locate the shoulder cut (**G**). Now repeat the process on the opposite face (**H**). Finally, fit the tenon into the mortise (**I**). If necessary, trim the tenon with a shoulder plane.

▶ See *"Mortise-and-Tenon Joint"* on p. 115 for information.

Dovetails on the Table Saw

Like the mortise-and-tenon, the dovetail is a strong joint that has broad applications. You can cut this joint entirely by hand, or you can cut the tail on the table saw and cut the socket with hand tools. This method speeds up the process yet still yields a "hand-cut" look.

➤ See *"Leg-to-Rail Dovetail"* on p.113.

Begin by laying out the tail and shoulder. Next, tilt the blade on your table saw to the required angle; in this example, it is 14 degrees. To guide the stock, attach a backup board to the miter gauge; a second, taller board clamped to the backup board registers each cut and supports the workpiece at 90 degrees **(A)**. Flip the piece over for the second cut **(B)**.

The next step is to cut the shoulders. Return the blade to 90 degrees, and reduce the height almost to the base of the shoulder. Because of the angle and the thickness of the blade, you can't cut all the way into the corner of the shoulder. Clamp a stop to the miter gauge to register the cut **(C)**. Each subsequent cut will align with the initial cut. Trim the inside corner of the shoulder with a sharp chisel.

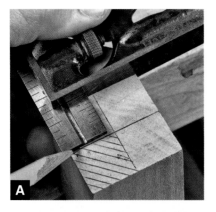

A

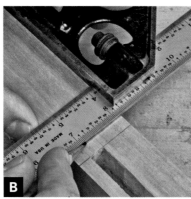

B

C

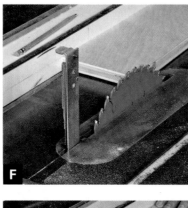

D

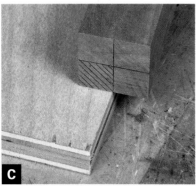

E

F

G

H

Tapering with the Table Saw

The simple tapered leg is found on a number of furniture styles, from Shaker to Federal. After tapering, you can quickly smooth the tapers with a bench plane.

▶ See *"Smoothing a Tapered Leg"* on p. 152.

Begin by laying out the taper. On the foot, draw the portion of the leg that will be sawn away (**A**). On the face of the leg, mark the start of the taper (**B**). Now use the layout lines to mark a jig. The jig is simply a piece of ¾-in. plywood that is long enough to encompass the leg and wide enough to position your hands a safe distance from the sawblade (at least 6 in.).

Position the leg over the plywood so that the layout overhangs at the foot (**C**) and the start of the taper (**D**). Now trace the leg and cut along the layout lines with a bandsaw to create a notch that the leg will fit within (**E**).

Now you're ready to saw the taper. Set the rip fence to the full width of the jig (**F**). Fit the leg into the notch and rip the taper on the first face (**G**); rotate the leg 90 degrees and rip the second taper on the adjacent face (**H**).

Cutting a Cove on the Table Saw

One of the most useful table-saw techniques is cove cutting. The concept is simple; by making a series of cuts at an angled approach, you can create an elliptical cove. This is an efficient yet inexpensive method for shaping almost any size of cove for moldings and furniture parts. The depth of the cove is determined by the blade height on the final pass, and the angle of the fence determines the width of the cove.

Begin by marking the width and depth of the cove on each end of the workpiece **(A)**. You can use a sawblade for cutting coves (a rip or combination blade works best), but this specially designed cove cutterhead from CMT leaves a much smoother surface that requires very little sanding **(B)**. Next, adjust the height of the cutter or blade to equal the depth of the cove **(C)**.

The next step is to position a fence at the correct angle **(D)**. Although finding the angle can seem difficult, it's really not. Using the layout on the stock as a guide, position the fence so that the cutter enters one edge of the cove **(E)** and exits at the opposite edge of the cove **(F)**.

(Text continues on p. 200.)

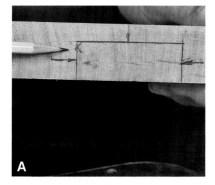

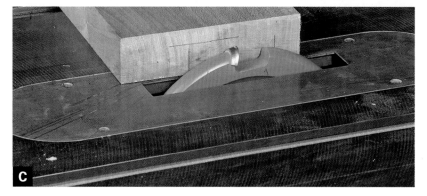

Slight angle to blade; nearly circular cove.

Moderate angle to blade; cove is narrower, slightly elliptical.

Steep angle to blade; cove is very narrow and elliptical.

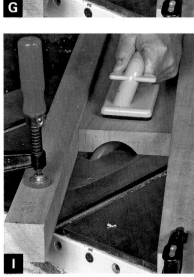

Woodworkers disagree about whether the fence should be positioned on the front or back of the blade. I prefer the back side because as I feed the workpiece, I'm pushing against the fence. With the fence in front of the blade, you'll inadvertently push the stock away from the fence and spoil the cut. However, to end the debate, you can use two fences; just position a second fence parallel to the first one (**G**). Use sturdy stock for the fences and clamp them firmly to the top of the table saw.

With the fences clamped into place, you're ready to cut the cove. Lower the cutter and take a series of light passes (**H**); use push blocks and raise the cutterhead approximately ⅟₁₆ in. each time (**I**). On the final pass, feed the stock very slowly to minimize sanding (**J**).

> ⚠ **WARNING** When cutting coves on the table saw, take very light cuts, especially on the final few passes.

The Jointer and Planer

Using the Jointer

Jointing and Planing

THE JOINTER AND PLANER WORK as a team to mill stock to final dimensions. The jointer is used to flatten the face of a board and straighten the edge. The planer mills boards to the finished thickness. Remember that a planer can't remove twist or warp from a board unless one face is flattened first.

The Jointer

The jointer is essentially an inverted plane that is used for truing faces and squaring edges when you're milling stock to size. The stock is first positioned on the infeed table and pushed across the cutterhead, where it is supported by the outfeed table. The two tables are parallel but not in the same plane;

lowering the infeed table with a handwheel or lever increases the cutting depth. After you joint one face and an edge, plane the workpiece to final thickness and rip to width using the jointed surfaces as reference surfaces.

Although many old jointers have powerful motors mounted directly to the shaft of the cutterhead, most jointers today are driven with a belt-and-pulley system. All jointers have a fence that supports the stock during edge jointing. As the knives become dull, the fence can be repositioned along the cutterhead to expose a less worn area. Although the fence tilts for beveling edges, using the jointer this way is awkward at best; the work has a tendency to slide down the fence and

THE JOINTER

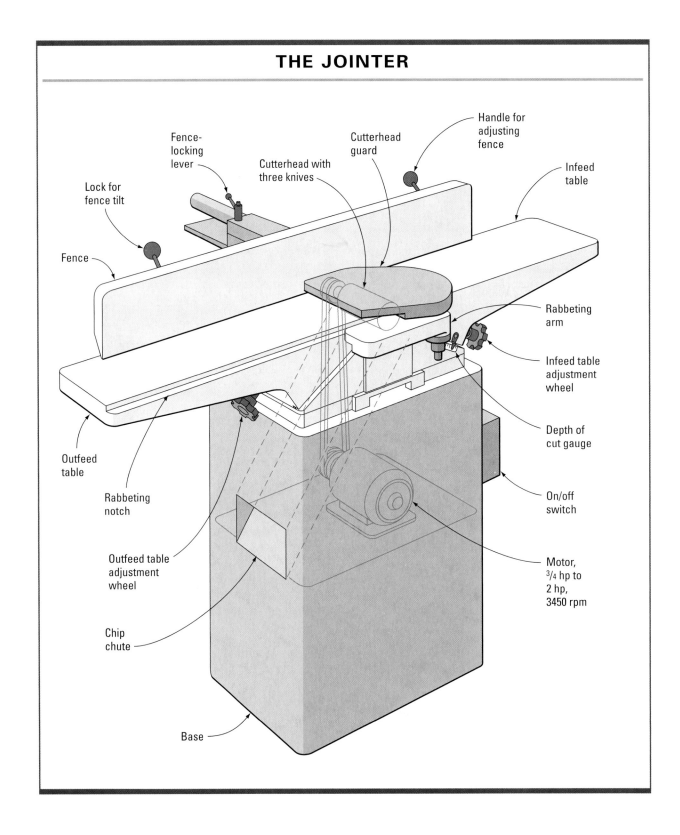

Fence-locking lever

Lock for fence tilt

Cutterhead with three knives

Cutterhead guard

Handle for adjusting fence

Infeed table

Fence

Outfeed table

Rabbeting notch

Outfeed table adjustment wheel

Chip chute

Base

Rabbeting arm

Infeed table adjustment wheel

Depth of cut gauge

On/off switch

Motor, $3/4$ hp to 2 hp, 3450 rpm

This extra-large 16-in. jointer has a direct-drive motor. Newer jointers have a belt-and-pulley drive system.

spoil the cutting angle. You'll find that it is much easier to rip the bevel on your table saw and smooth the sawn surface with a bench plane.

Jointer Size

Six-in. jointers are common, but if you plan to use wide lumber, you'll want a jointer that is closer to the size of your planer. Remember, it's important to flatten stock on the jointer before planing it to thickness. If your budget allows, shop around for an 8-in. or even a 12-in. machine. If not, then you can always flatten wide boards with a bench plane.

Jointer Tune-Up

Unlike a planer, a jointer does not need many adjustments. For smooth, accurate cuts, it's important to keep the knives sharp. When installed, either new or after sharpening, the knives must be set at the same height in the cutterhead. A magnetic

> ### ► JOINTER SAFETY

All new jointers come equipped from the factory with a guard; unless you are changing the knives, there is absolutely no reason to remove it. The guard pivots out of the way to expose the cutterhead as the stock is pushed through. Afterwards, a spring swings it back into position to cover the spinning knives.

Here are additional safety guidelines:

• Always read and follow manufacturer's instructions.

• Use push sticks and push blocks to distance your hands from the cutterhead.

• Never joint stock less than 12 in. long.

Setting all the jointer knives at the same height is crucial to the machine's operation. A magnetic jig makes the task easier.

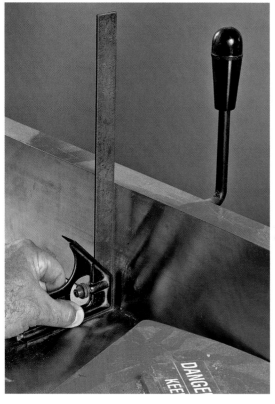

Use a good combination square or machinist's square to set the jointer fence square to the table.

JOINTER TABLE ALIGNMENT

Infeed and outfeed tables must be parallel with each other and with the cutterhead.

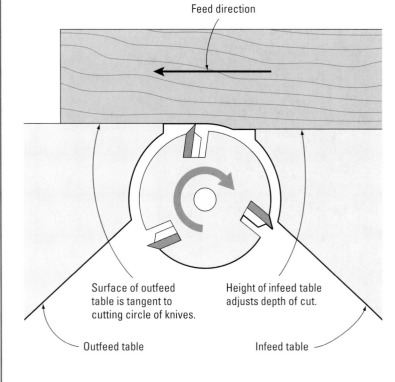

Feed direction

Surface of outfeed table is tangent to cutting circle of knives.

Height of infeed table adjusts depth of cut.

Outfeed table

Infeed table

knife-setting device will make this job much easier. The other important adjustment is the outfeed table height: It should be set even with (or tangent to) the cutting arc of the knives.

After setting new knives in the cutterhead, you may need to make a slight adjustment to the outfeed table. If too low, the knives will snipe the trailing end of the board as the cut is completed. If the table is too high, it will cut a progressively convex face or edge. Once the outfeed table is set at the correct height, lock it tightly. You should never have to adjust it until you sharpen or change the knives. Finally, check the fence for squareness to the table.

The Planer

The planer smoothes and sizes stock to final thickness. Although you can do this chore by hand with a bench plane, it makes sense to reduce this labor-intensive task to a few minutes with a power planer. However, it's important to remember that a planer does not remove warp in a board, so you must first flatten the stock with a jointer or bench plane.

Not too many years ago, planers were bulky, heavy, and expensive, which positioned them out of reach for many woodworkers. Fortunately, today there are a number of 12-in.-to-15-in. planers that are rugged, compact, and affordable. Also, don't overlook the popular benchtop models; they are very lightweight and truly portable machines but still do an effective planing job.

A planer is a work-horse that efficiently smoothes rough stock.

Benchtop planers are lightweight, compact, and affordable.

To effectively tune up your plane, you'll need a dial indicator.

How a Planer Works

With their system of feed rollers, bed rollers, pressure bar, chipbreaker, and cutterhead, planers are among the most complex of woodworking machines. Each part plays an important role in the difficult task of smoothing a board to a uniform and precise thickness. Tuning is important, too. During tune-up, each part of a planer is precisely adjusted in relationship to the bed or cutterhead. Fortunately, tuning a planer isn't difficult once you take a few minutes to familiarize yourself with the various parts and their functions.

THE PLANER

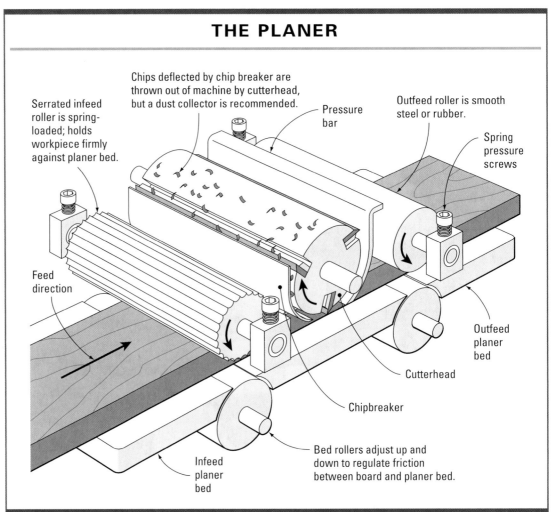

Chips deflected by chip breaker are thrown out of machine by cutterhead, but a dust collector is recommended.

Pressure bar

Outfeed roller is smooth steel or rubber.

Spring pressure screws

Serrated infeed roller is spring-loaded; holds workpiece firmly against planer bed.

Feed direction

Outfeed planer bed

Cutterhead

Chipbreaker

Infeed planer bed

Bed rollers adjust up and down to regulate friction between board and planer bed.

The cutterhead typically holds three knives, but large industrial planers may have four, while benchtop models usually have two. For a smooth surface you'll need to change the knives occasionally, depending on how often you use the machine. A cast-iron table supports the work as feed rollers move it through the machine. The infeed roller is serrated to provide a firm grip on rough stock. To avoid marring the freshly planed surface, the outfeed roller is smoothly machined steel or coated with rubber. Bed rollers, located below the feed rollers, help reduce friction as a board makes its short trip through the machine.

As each knife lifts a shaving, the spring-loaded chipbreaker applies pressure to the board to "break" the chip, which helps limit tearout. The curved contour of the chipbreaker also serves to deflect the chips toward the dust collector. The pressure bar is also spring-loaded and helps prevent "chatter," or vibration of the board, as each knife contacts the stock.

Power is supplied from the motor to the cutterhead and feed rollers with a belt-and-pulley system. The rpm are reduced to the feed rollers through gear reduction.

The best benchtop planers feature a cutterhead lock to prevent sniping the end of the stock. And don't forget the dust-collection hood. If it's not standard equipment with your machine, it is an accessory worth purchasing or making yourself.

Most benchtop planers feature a cutterhead lock to minimize sniping.

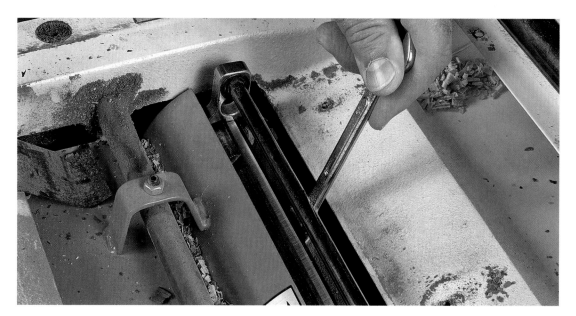

Planer knives must be replaced once they start to become worn and dull.

TROUBLESHOOTING PLANER PROBLEMS

Although there is a lot that can go wrong with a planer, most problems are easy to analyze and repair. Here's a list of common problems and solutions.

PROBLEM	CAUSE	SOLUTION
Ridges in stock	Chip in knives	1. Shift one knife to left or right. 2. Sharpen knives.
Stock hangs up inside planer	1. Pressure bar too low	1. Adjust height of pressure bar.
	2. Bed rollers too low	2. Adjust height of bed rollers.
	3. Insufficient pressure from feed rollers	3. Increase spring tension in feed rollers. Lower feed rollers.
Surface of stock burned or glazed	Knives are very dull	Sharpen knives.
Ridges in stock are rough and irregular	Pressure bar isn't holding stock to table	Increase spring tension or lower the pressure bar.
Board turns diagonally in planer	Uneven feed roll pressure	Adjust feed roll.

Dust collection is a must with any planer.

Despite appearances, a freshly planed board isn't truly flat but is a series of ridges that are cut as each knife takes a "bite" from the board. The spacing of the ridges varies with cutterhead speed and feed rate. Slower feed rates and/or higher cutterhead speeds will reduce the space between each ridge. Some planers have two speeds for the feed rate—a faster feed rate for efficient planing, and a slower rate for a smoother surface. I opt for the faster rate. Because I always "finish" each board by smoothing the surface with a bench plane, ridge spacing is not critical.

Face-Jointing

Rough lumber should be flattened before it's planed to thickness to remove the warp. Remember, a thickness planer doesn't remove warp; it just smoothes the surface and reduces the thickness. The jointer is like a large motorized bench plane. Because the planer and jointer work as a team, it's best to have a jointer that matches the size of your planer.

Begin by setting the cutting depth of the jointer to ½₂ in. Check along the edge and end of the board to determine whether it's bowed, which is typical. Always joint the face of a board that is concave along its length; if you try to joint the convex face, the board will follow the arc as it's fed across the tables and will never become flat. Also, check the grain direction of the board to determine the best way to feed it past the cutter **(A)**. Position the board on the infeed table and use push blocks to hold it down against the table **(B)**. Use a push block to feed and guide the work across the cutterhead **(C)**. As the stock passes the cutterhead, transfer all the downward pressure to the outfeed table as you push **(D)**. If you allow the stock to rise from the outfeed table, even slightly, it will be convex rather than straight. After the first pass, you'll notice that wood has been cut away at each end of the stock as the jointer cuts a straight path **(E)**. One or two more additional passes across the jointer will complete the job **(F)**.

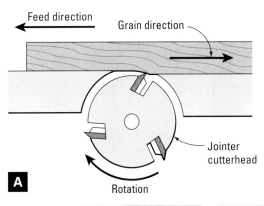

A — Feed direction · Grain direction · Jointer cutterhead · Rotation

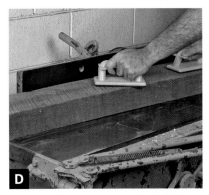

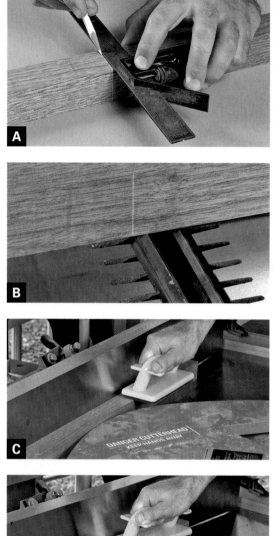

A

B

C

D

E

Tapering with a Jointer

You can cut a taper on your jointer by taking several passes from the start of the taper. Begin by marking a line at the start of the taper **(A)**. Next, set the cutting depth to ⅟₁₆ in. and position the start of the taper just slightly beyond the arc of the cutterhead **(B)**. Now clamp a stop block to the infeed table; position the stop block against the end of the stock.

With the setup complete, you're ready to cut the taper. Position the stock against the stop block and carefully lower it into the spinning cutterhead. Once the stock contacts the table, feed it forward with push blocks **(C)**. Take several passes at ⅟₁₆-in. deep until the full depth of the taper is reached **(D)**. Repeat the process for each tapered face **(E)**.

> ⚠ **WARNING** Never let your hand trail the back of the workpiece on the jointer. Always use a push block, especially to guide the back end of the workpiece.

Milling Square Stock

Square stock is used often in furniture making for such items as table legs and bedposts. If you're starting with a wide plank, rip it down first on the bandsaw.

➤ See *"Ripping with a Fence"* on p. 232.

The first step in jointing is to flatten a face. Afterwards, an adjacent face is jointed straight and 90 degrees to the first face. It's easier to flatten a concave face; it has two surfaces resting on the jointer table **(A)** instead of just one **(B)**. Position a push block at each end and take a light cut **(C)**. You'll notice that the jointer will cut stock from each end of the workpiece **(D)**. Now take one or two more passes until the workpiece is smooth and flat along the entire length **(E)**.

Now you're ready to square an adjacent face. First, check the jointer fence for squareness to the table and adjust it if necessary **(F)**. Next, position the workpiece on the jointer with the

(Text continues on p. 212.)

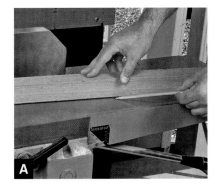

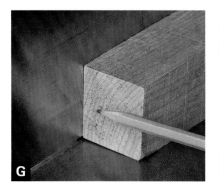

smooth, true face against the fence **(G)**. It's important to view the workpiece from the end; although it may appear to be against the fence when viewed from the top, it may not be making full contact **(H)**. Now joint the second face **(I)** and check the two surfaces for square. **(J)**

Now you're ready to plane the stock to final thickness **(K)**. Always position the jointed face down and remember to plane both the remaining two surfaces **(L)**. Finally, square the end **(M)** and use a stop to cut the piece to final length **(N)**.

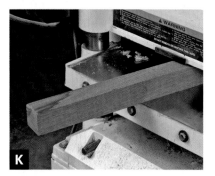

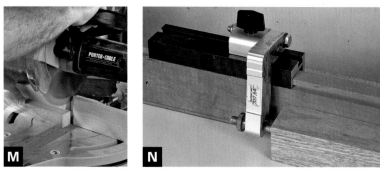

Milling Flat Stock

This process is used for sizing flat panels such as drawer fronts, door panels, and tabletops. Begin by cutting the stock to a width and length that is approximately ½ in. greater than the final width and length. Reducing the stock size makes it easier and more efficient to work. However, don't cut the stock shorter than 12 in. because it's unsafe to joint or plane at that length.

Next, position the concave face of the stock on the jointer **(A)**. The convex face (opposite face) will have a tendency to rock, which makes it more difficult to flatten **(B)**. Set the jointer for a light cut and use push blocks to feed the stock across the cutterhead **(C)**. As you push the stock, keep it firmly on the outfeed table **(D)**. For safety, continue pushing the stock past the guard until it closes against the fence **(E)**. The first pass will reveal the low spots on the board **(F)**. Now position the smooth, flat face down on the bed of the

(Text continues on p. 214.)

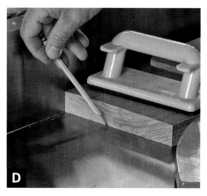
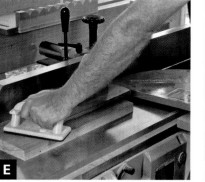

G

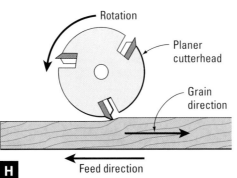

Rotation

Planer
cutterhead

Grain
direction

Feed direction

H

planer and plane the stock to thickness **(G)**. Always check the grain direction of the board and feed the stock accordingly **(H)**.

Next, joint one edge of the stock. As you joint the edge, keep the face firmly against the jointer fence **(I)**. Now rip the board to final width using the jointed edge as a guide against the fence **(J)**. Cut the first end 90 degrees to the edge **(K)**, then use a stop to cut the board to final length **(L)**.

I

J

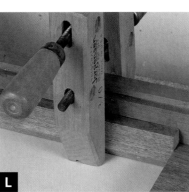

K

L

The Bandsaw

Tune-Up Techniques

➤ Changing and Tensioning Blades (pg. 223)

Cutting Simple Curves

➤ Bandsawing Tight Curves (p. 225)

➤ Bandsawing Broad Curves (p. 227)

Compound Curves

➤ Cabriole Leg (p. 228)

➤ Ogee Foot (p. 230)

Bandsaw Ripping

➤ Ripping Freehand (p. 231)

➤ Ripping on the Bandsaw with a Fence (p. 232)

➤ Resawing (p. 233)

Bandsaw Joinery

➤ Tenons on the Bandsaw (p. 235)

L IKE THE TABLE SAW, the bandsaw is a universal machine capable of performing many woodworking tasks. But unlike a table saw, the bandsaw can't cause a kickback—when the workpiece is thrown back at the operator by the blade. Also, the kerf of a bandsaw blade is only half as wide as a kerf from a table saw. This makes the bandsaw ideal for ripping heavy warped stock that would bind and bog down a table saw. Of course, the bandsaw excels at sawing curves of almost any form—broad, tight, or compound. And the bandsaw is the only machine that can resaw, which is the process of ripping a thick board into thinner ones. If you've ever wanted to create bookmatched door panels or saw your own veneer from a prized plank, you'll appreciate the unique features of a bandsaw.

The bandsaw's versatility stems from its thin, flexible blade, which is stretched around two or three wheels. Blade sizes range from $1/16$ in. for cutting intricate scrollwork to wide blades, to $3/4$ in. or more, used for resawing and cutting broad, sweeping curves. More on blades in a minute. Let's first look at the other components of this versatile tool.

Bandsaw Anatomy

The wheels of a bandsaw support the blade and transmit power from the motor. To cushion the blade, provide traction, and protect the sharp teeth, wheels are fitted with

THE PARTS OF A BANDSAW

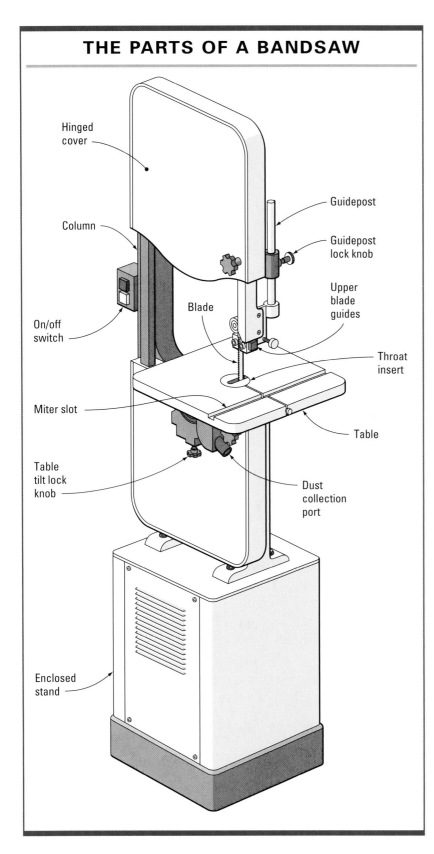

Hinged cover

Column

On/off switch

Blade

Miter slot

Table tilt lock knob

Guidepost

Guidepost lock knob

Upper blade guides

Throat insert

Table

Dust collection port

Enclosed stand

BANDSAW BACK VIEW

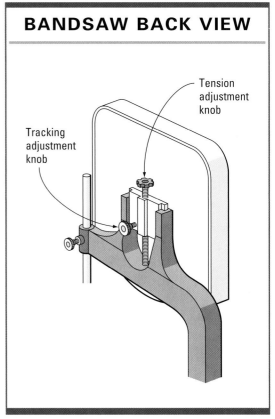

Tension adjustment knob

Tracking adjustment knob

tires. Tires are usually crowned in profile on smaller bandsaws, which makes it easier to track the blade. Large bandsaws typically have flat tires that give better support to blades over 1 in. wide.

The diameter of the wheels determines the throat size, or the distance from the blade to the column. Three-wheel bandsaws have a larger throat capacity because the wheels form a triangle. But the small-diameter wheels on three-wheel saws tend to break blades frequently as they flex the blade sharply through each revolution of the saw. Also, because of the inherent flex of the frame, three-wheel saws cannot adequately tension most types of blades.

The distance between the wheels determines a bandsaw's resaw capacity. For woodworkers who plan to do a lot of resawing, this is one of the most important features. And not all brands of bandsaws with the same wheel diameter have equivalent resawing capacities. So it's smart to check this important dimension when you're shopping for a saw. Most 14-in. bandsaws accept an accessory riser block that bolts into the frame to double the resaw capacity. This feature, along with the compact size and attractive price, makes 14-in. bandsaws a favorite in many home and professional shops.

To support the blade through the twists and turns of cutting curves, bandsaws come equipped with a pair of guides. The guides use either blocks or rollers to support the blade on each side. Thrust wheels support the blade from behind to keep it from being pushed off the wheels. Having the guides tuned and adjusted is key to the top performance of your bandsaw.

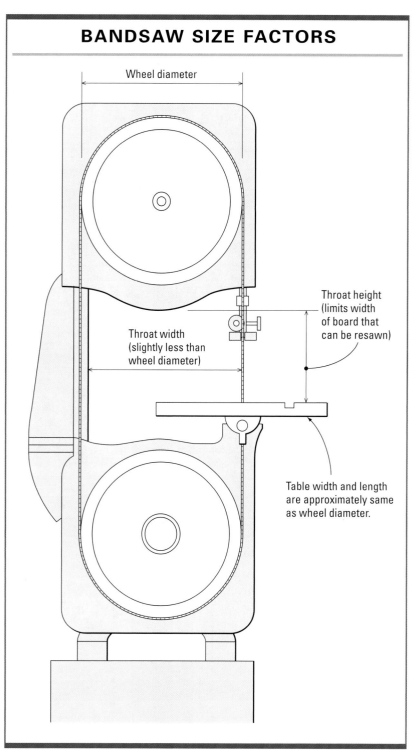

BANDSAW SIZE FACTORS

Wheel diameter

Throat height (limits width of board that can be resawn)

Throat width (slightly less than wheel diameter)

Table width and length are approximately same as wheel diameter.

Bandsaw guides provide essential support to the blade throughout the cut. These are the upper guides; an identical pair is mounted below the table.

➤ BANDSAW GUIDES

Open most any woodworking catalog, and you're sure to see pricey aftermarket guides for your bandsaw. The idea is often touted that the steel guide blocks that are standard on most 14-in. bandsaws are inferior. With plastic blocks or ball-bearing guides, you're promised less heat, longer blade life, and better blade support. But the fact is that the steel block guides on your saw can work as well or better than aftermarket guides. First of all, block guides have a broader surface area than bearings and do a better job at limiting blade flex through the twists and turns of curve cutting.

Also, despite what you've heard, block guides don't overheat the blade and shorten its life. The heat is caused by the friction of cutting, just as your thumbs get hot when pushing a scraper. Instead of expensive guides, purchase a bi-metal (high-speed steel) or carbide-tipped blade, either of which is designed to handle the heat and rigors of resawing.

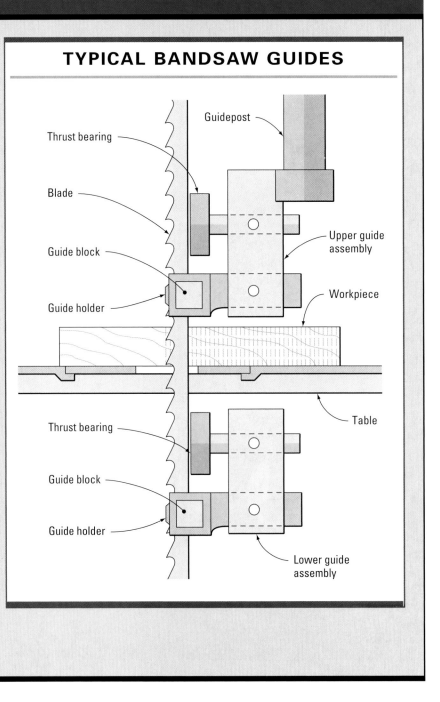

TYPICAL BANDSAW GUIDES

Thrust bearing
Blade
Guide block
Guide holder
Guidepost
Upper guide assembly
Workpiece
Table
Thrust bearing
Guide block
Guide holder
Lower guide assembly

The bandsaw is the best choice for ripping thick stock.

The table supports the stock during cutting. Also, most tables tilt for angled cuts. Most bandsaw tables are milled with a miter-gauge slot, which allows you to make surprisingly accurate crosscuts. Many bandsaws also come equipped with a fence, which makes ripping accurate and effective.

The upper guide of a bandsaw is attached to a post that adjusts vertically for various thicknesses of stock. Attached to the post is a guard that serves as a barrier to shield your hands from the moving blade. For your personal safety, it's important to lower the guidepost before turning on the saw. Position the stock next to the blade and adjust the guide so it is approximately ¼ in. above the stock.

The wheel covers are another important safety feature. Always keep the covers in place unless you are changing blades. To make changing blades more efficient, I prefer hinged covers and a quick-release catch. A large tension handwheel also makes it easier to change blades.

This tension wheel is large and positioned below the bandsaw wheel, making it easy to reach.

Bandsaw Blades

The blade is the heart of any bandsaw. Even the finest bandsaw will perform poorly with the wrong blade or a dull blade. Unlike that on a table saw, there is really no "combination" bandsaw blade that will work for a wide variety of cuts. To get the best per-formance from your saw, it's best to change blades. If you peruse the pages of a bandsaw blade catalog, you'll see a dizzying assort-ment of blade types. But understanding the basics will help you sort through the choices.

Blade Pitch

Pitch is the number of teeth per inch (or tpi). As on any saw, a greater number of teeth will give a smoother, slower cut while fewer teeth will give a faster, rougher cut. A good rule of thumb is to select a blade that will place no fewer than 6 and no greater than 12 teeth in the stock at one time. For example, either a 6- or 10-tpi blade would be a good choice for cutting 1-in. stock.

Tooth Form

Most blade retailers offer three types of tooth forms to choose from—regular, skip, and hook. Regular-tooth blades have a 0-degree rake angle, which yields a smooth cut. (Rake is the angle of the tooth relative to the blade body.) This blade is a good choice for smooth sawing of curves in stock 1-in. thick or less.

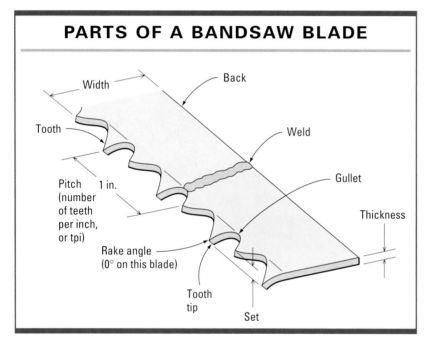

PARTS OF A BANDSAW BLADE

- Width
- Back
- Tooth
- Weld
- Pitch (number of teeth per inch, or tpi)
- 1 in.
- Gullet
- Thickness
- Rake angle (0° on this blade)
- Tooth tip
- Set

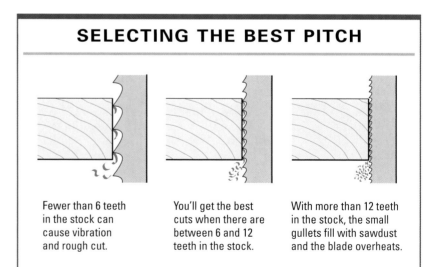

SELECTING THE BEST PITCH

Fewer than 6 teeth in the stock can cause vibration and rough cut.

You'll get the best cuts when there are between 6 and 12 teeth in the stock.

With more than 12 teeth in the stock, the small gullets fill with sawdust and the blade overheats.

Like the regular-tooth form, the skip tooth has a smooth-cutting 0-degree rake angle. As the name implies, every other tooth is omitted, so there are half the teeth per in. of a regular-tooth blade. This configuration creates a large gullet and places fewer teeth in contact with the stock—ideal for sawing thick planks.

However, a better choice for resawing and sawing thick stock is the hook tooth. Similar to the skip-tooth design, this one also features large teeth and gullets, but its blades also have a positive rake angle that cuts more aggressively than a skip-tooth blade, and with less feed resistance. Hook-tooth blades are a great choice for resawing and ripping thick stock.

Still another option is the variable-pitch blade. With this unique design, the teeth vary in size but not in shape. Each tooth is a hook style, but the variable pitch minimizes harmonic vibrations, which helps improve the smoothness of the cut. Variable-pitch

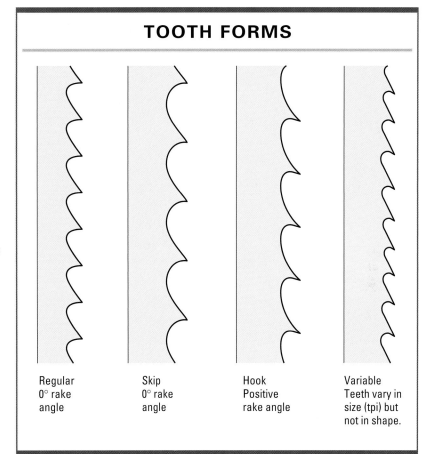

TOOTH FORMS

Regular
0° rake
angle

Skip
0° rake
angle

Hook
Positive
rake angle

Variable
Teeth vary in
size (tpi) but
not in shape.

▶ BANDSAW SAFETY

• Always follow the manufacturer's safety guidelines for your saw.

• Keep your fingers out of the path of the blade.

• Decrease feed pressure as you approach the end of the cut.

• Use push sticks when ripping or resawing.

• Keep the wheel covers shut when the saw is running.

• Adjust the upper guide approximately ¼ in. above the workpiece before starting the saw.

• Keep the blade guard in place.

• Disconnect saw from power source before changing blades.

• Always wear eye protection when working with the bandsaw.

blades are ideal for resawing, especially veneer. They cut quickly and aggressively yet leave a fine surface that requires little additional smoothing and cleanup.

Blade Width

Blade width ranges from $1/16$-in. blades for scrolling to large 2-in. blades that cut a straight path for ripping or resawing. Selecting the best blade width isn't difficult, but there are a few things to keep in mind.

First, as blades increase in width, they also increase in thickness. A thick, wide blade requires more beam strength to tension than a narrow, thin one. Also, thick blades need large-diameter wheels to prevent excessive bending stress and premature breakage. So check the manual with your bandsaw, and don't exceed the maximum blade width. In fact, although conventional wisdom states that you'll get a truer cut from

a wide blade, the frames of many consumer bandsaws lack sufficient stiffness to adequately tension a wide blade, despite the tension-scale reading on the saw. When you're resawing, your bandsaw will cut a straighter path through that prized plank using a narrow blade under sufficient tension than with a wide blade insufficiently tensioned. So although most 14-in. bandsaws can accept a $3/4$-in.-wide blade, you'll get better results resawing with a $3/8$-in. or $1/2$-in. blade. A great choice for resawing is a $3/8$-in. or $1/2$-in. variable-pitch hook-tooth blade.

When sawing curves, choose the widest blade that will fit the curve. Obviously a $1/4$-in. blade will follow curves of most any size, but it will have a tendency to wander if used for sawing broad curves. You'll get a more accurate outline and spend less time fairing broad curves if you switch to a wider blade.

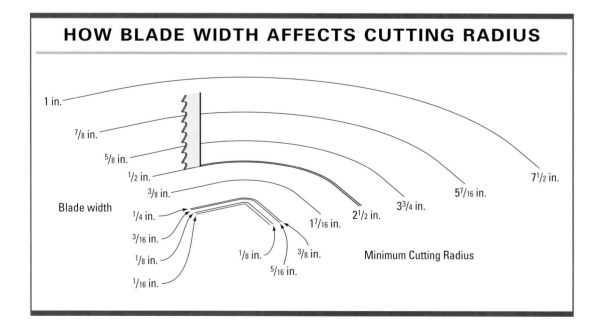

HOW BLADE WIDTH AFFECTS CUTTING RADIUS

Changing and Tensioning Blades

You'll get much greater versatility from your bandsaw if you take a few minutes to change the blade as the need arises. For example, using a wide blade to make tight curves will just bend the blade, put stress on the guides, and burn the wood. And attempting to resaw with a dull blade will bring nothing but disappointment.

► For information on blade selection, see p. 220.

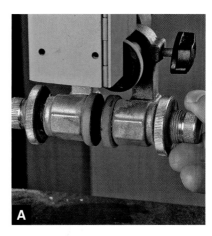

To change the blade on your bandsaw, begin by disconnecting the saw from its power source. Next, loosen the tension, remove the old blade, and open the blade guides **(A)**. Now begin installing the new blade by hanging it over the upper wheel, then fit it around the lower wheel. Apply just enough tension to remove the slack from the blade.

The next step is to track the blade. While spinning the upper wheel with your left hand, slowly rotate the tracking knob with your right hand. Watch the blade as it rides the upper wheel and turn the tracking knob slowly to keep it centered on the tire **(B)**. Once the blade is tracking well, increase the tension. Realize that you may have to make slight adjustments to the tracking as the tension is increased.

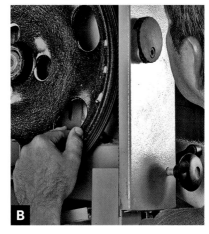

For sawing curves in relatively thin stock, it is usually enough just to remove the slack from the blade until it is taut. However, when you're tensioning a blade for resawing, it's best to use a tension meter **(C)**. If you don't own a tension meter, try raising the upper guide all the way and

(Text continues on p. 224.)

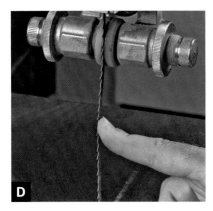

The thrust bearings and guides surround the blade and keep it from bowing, twisting, or wandering in a cut. Adjust them so they are not quite touching the blade when idle.

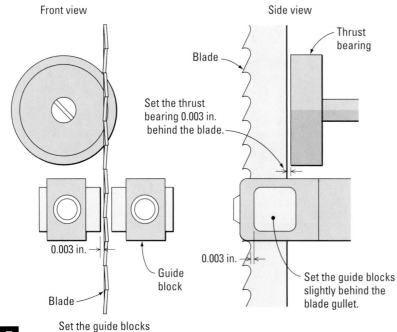

Front view

Side view

Blade

Thrust bearing

Set the thrust bearing 0.003 in. behind the blade.

0.003 in.

Guide block

Blade

0.003 in.

Set the guide blocks slightly behind the blade gullet.

Set the guide blocks 0.003 in. from the blade.

E

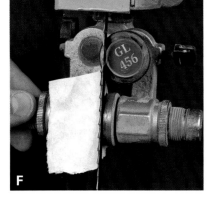

deflecting the blade sideways with your index finger—obviously with the saw turned off and disconnected from the power source **(D)**! With a moderate amount of finger pressure, the blade should deflect no more than ¼ in. Obviously, this method isn't nearly as precise as using a gauge, but with a little experience it works well.

The next step is to adjust the guides **(E)**. A scrap of paper or even a dollar bill works nicely as a feeler gauge **(F)**. Adjust each side guide evenly to avoid sideways blade deflection. Also, remember to adjust the thrust wheels behind the blade using the paper feeler gauge. The thrust wheels play the important role of preventing the blade from being pushed off the wheels during sawing.

The last step is to check the squareness of the table to the blade **(G)**. This step is critical if you are planning on resawing wide boards. If necessary, adjust the trunnions under the table to bring it into square.

Bandsawing Tight Curves

Bandsawing tight, decorative contours is one of the most common uses for the bandsaw. First, select a narrow blade that will fit through the contours of the pattern. Next, trace the pattern on the stock **(A)** and adjust the upper blade guide approximately ¼ in. above the workpiece **(B)**.

Before cutting curves, it's important to plan the sequence of cuts. You'll want to avoid backing up through a curve; the kerf will bind on the blade and pull it off the wheels. Instead, try to avoid trapping the blade in a corner, and if you must back up, plan to back out of straight cuts rather than curves.

Begin by cutting inside corners of the profile **(C)**. This will create relief cuts. Also, look at the pattern for curves that don't end at an inside corner and saw these first **(D)**. As you approach a tight curve, slow down and carefully follow the outline **(E)** to avoid miscuts and excess handwork later on.

(Text continues on p. 226.)

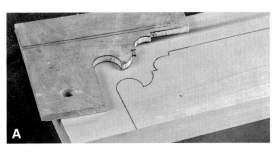

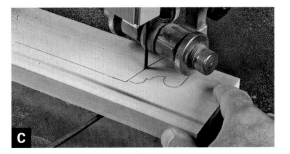

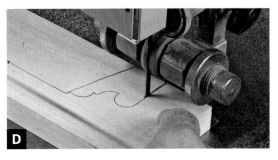

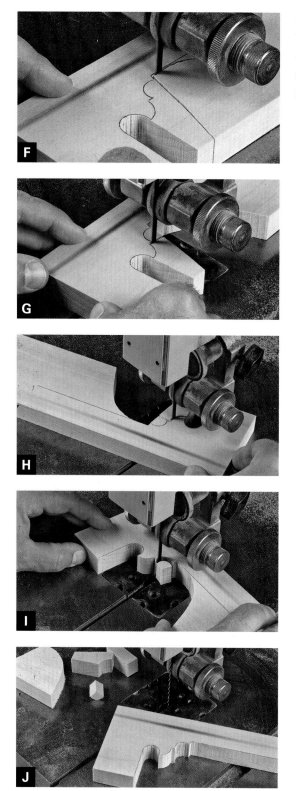

Continue sawing the outer portions of the pattern to gain access to the inner areas of the design **(F)**. A short, straight relief cut **(G)**, followed by short convex **(H)** and concave curves **(I)**, completes the outline **(J)**.

Bandsawing Broad Curves

Broad, sweeping curves can be sawn with a narrow blade, but you'll find that you'll saw more accurately and efficiently if you switch to the widest blade possible for the curves—a ½-in.-wide blade, in this case **(A)**. As you saw the contours, position your hands on each side of the workpiece for the greatest control **(B)**. The end result will be a fair curve **(C)** that needs only minor smoothing with a spokeshave.

[**TIP**] When sawing curves, lay out a crisp line. Saw on the waste side of the line, then smooth the wood right to the layout line.

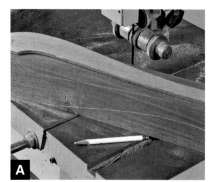

A

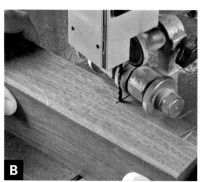

B

Cabriole Leg

Compound curves are those that curve in two directions simultaneously. Probably the most familiar example is the cabriole leg. But many examples of contemporary furniture use compound curves as well. The technique for creating compound curves isn't difficult. In fact, it takes only a few minutes to saw most compound curves, and the results are exciting.

Start with a pattern of the curves. Trace the pattern onto two adjacent surfaces of the stock (**A**). Realize, too, that some compound curves are asymmetrical, so they will use two different patterns, one for each adjacent face.

Now you're ready to begin sawing. Because of the long length of the leg in this example, I begin by sawing a "bridge" (**B**), just an area of support that remains until the end of the second sequence of cuts (**C**). Two short parallel cuts create the bridge (**D**). Next, start sawing the curve at the foot (**E**), and follow the outline until it reaches the bridge (**F**). Now turn the leg stock end-to-end and start a cut

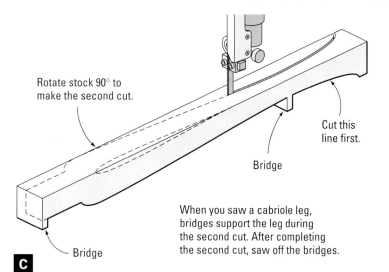

Rotate stock 90° to make the second cut.

Cut this line first.

Bridge

Bridge

When you saw a cabriole leg, bridges support the leg during the second cut. After completing the second cut, saw off the bridges.

C

D

E

F

at the knee **(G)**. Saw slowly and carefully to minimize handwork when smoothing the curve **(H)**.

Once the front curves are complete, turn your attention to the curved cut at the back of the leg **(I)**. Save this offcut and tape it back into position **(J)**; it has the layout for the adjacent curves.

Next, turn the leg 90 degrees and complete the second sequence of cuts **(K)**. Finally, cut off the bridge **(L)**.

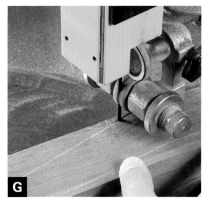

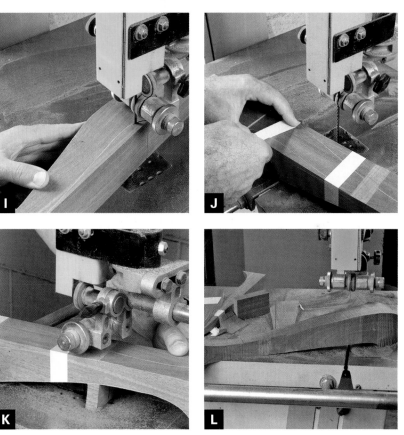

A

B

C

D

Ogee Foot

The ogee foot is a somewhat different example of a compound curve; instead of sawing two surfaces on one piece of stock, first bandsaw two pieces of stock individually, join them with a miter, and then bandsaw them as an assembly.

Begin with patterns for the bracket outline as well as the ogee contour. After cutting the miter and bandsawing the bracket outline, glue the two halves together to create a foot.

Position the foot on a support stand, and clamp it into place **(A)**. Now you're ready to bandsaw the ogee contour, but first check the table and blade for square **(B)**. Slowly and carefully bandsaw the first face **(C)**. The outline for sawing the second face will be revealed in the miter **(D)**. Saw along the outside of the miter line, then sand the surfaces smooth.

Ripping Freehand on the Bandsaw

Ripping thick, rough stock is much safer and more efficient on the bandsaw than on the table saw. Unlike a table saw, the bandsaw can't kick back. And the thin blade of the bandsaw won't bog down in thick stock.

Begin by marking a layout line **(A)**. Select a ½-in. or wider blade and adjust the upper guide to the thickness of the stock **(B)**. As you rip the stock, you may have to adjust the angle of the work slightly to compensate for blade drift **(C)**. Position your hands on each edge of the stock for greatest control **(D)**. As you exit the stock, reduce the feed pressure, and keep your thumbs out of the blade's path **(E)**.

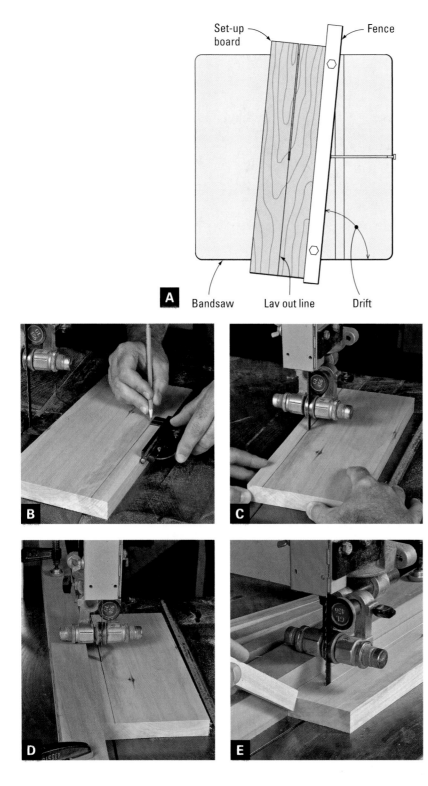

A Set-up board · Fence · Bandsaw · Lay out line · Drift

Ripping on the Bandsaw with a Fence

The bandsaw can make surprisingly accurate rip-cuts. The key is to use a fence. However, some bandsaws don't rip in a path that is square to the table, a phenomenon called drift. The solution is to angle a wood fence to compensate for the drift **(A)**.

Start by marking a line parallel to the edge of the board **(B)**. Next, use a ½-in.-wide blade and cut freehand along the line; as you carefully follow the line, you will naturally feed the stock at an angle to compensate for the drift **(C)**. About midway down the board, stop the saw and clamp a board along the edge of the stock to serve as a fence **(D)**. As you continue to rip the stock, distance your fingers from the blade by using a push stick **(E)**.

Resawing

Resawing is the process of ripping a thick board into thinner ones. You can use this technique for creating book-matched panels **(A)**, matching sheets of veneer, or just saving lumber when building small projects. It's a great technique that can only be done on the bandsaw. But first you'll need to select a blade for resawing. For this demonstration I'm using a ⅜-in., 3-tpi hook-tooth blade **(B)**.

[TIP] If you own one of the many 14-in. bandsaws on the market, you can double the resaw capacity, from 6 in. to 12 in., with a riser block.

To resaw accurately, it's best to use a tall fence to provide support to the entire width of the workpiece. You can attach a wide board to the rip fence on your bandsaw, or you can make a fence from plywood **(C)**. Begin the process by setting the fence at the drift angle. Mark a stick to the width that you want to resaw **(D)** and cut free-hand along the line **(E)**. When you reach the halfway point along the length of the stick, stop

(Text continues on p. 234.)

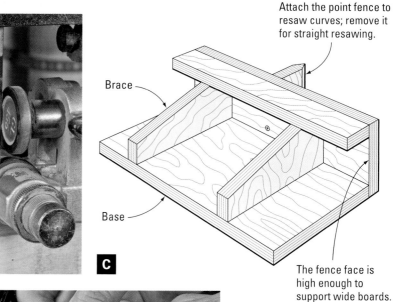

Attach the point fence to resaw curves; remove it for straight resawing.

Brace

Base

The fence face is high enough to support wide boards.

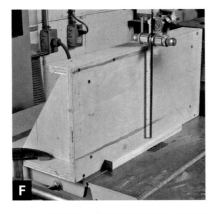

the saw. You will have naturally compensated for the drift angle, if any, as you follow the line. Now simply position the rip fence next to the stick and secure it to the table with clamps **(F)**. Now you're ready to begin resawing.

As you start the cut, keep the stock positioned firmly against the fence **(G)**. Feed the workpiece slowly, be aware of the sounds and vibrations of the bandsaw, and adjust your feed rate if necessary **(H)**. As you approach the end of the stock reduce the feed pressure and use a block of wood to complete the cut **(I)**.

Tenons on the Bandsaw

You can cut accurate tenons on the bandsaw if you use a wide blade (½ in. works well) and a fence to guide the cut. As always, cut the mortise first, and then cut the tenon to fit. If you need to adjust the fit, it is always easier and more accurate to adjust the tenon.

Begin by laying out the face and shoulder of the tenon (**A**). Then set the fence in position to cut along the layout line of the tenon face (**B**). Now turn on the saw and make the cut (**C**). A stick clamped to the fence works as a stop and makes cutting multiple tenons fast and accurate (**D**).

The next step is to cut the shoulder. Remove the fence and use a miter gauge to accurately guide the cut (**E**). Then rotate the workpiece and cut the second shoulder (**F**) to complete the tenon (**G**).

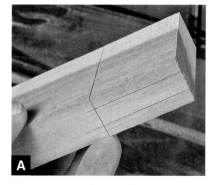
A

B

C

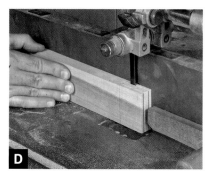
D

E

F

G

The Shaper

Shaping Edges

➤ Shaping Part of an
Edge (p. 242)

➤ Shaping the Entire
Edge (p. 243)

➤ Raised-Panel
Shaping (p. 244)

Shaping a Face

➤ Shaping a Curved
Casing (p. 246)

IF YOU'VE EVER USED a table-mounted
router to shape a simple molding or
bevel the edges of a door panel, then, in
effect, you've used a shaper. Many of the features, cutter profiles, and even the techniques are the same. Still, the largest router
is no match for the size and horsepower of a
shaper. Heavy cuts, such as in shaping raised
panels, usually require several passes with the
router, but the shaper easily breezes through
heavy cuts in a single pass.

Shaper Anatomy

Compared to most other woodworking
machines, the shaper is quite simple in
design—a heavy cast-iron top with a vertical
spindle protruding through a hole in the
center. Below the table is a powerful induction motor that, unlike the universal motor
in a router, is capable of delivering its rated
horsepower all day long. The motor produces horsepower through a belt-and-pulley
setup. Most shapers have a double pulley
arrangement that provides two speeds, typically 7,000 and 10,000 rpm. Multiple speeds
are important because larger-diameter cutters should be run at a slower speed than
smaller-diameter cutters.

The spindle is the heart of the shaper.
While router collets are either $\frac{1}{4}$ in. or
$\frac{1}{2}$ in., shaper spindles range in size from
$\frac{1}{2}$ in. in diameter to $1\frac{1}{4}$ in. The large size
of shaper spindles, coupled with the raw
horsepower of an induction motor, is the

THE SHAPER

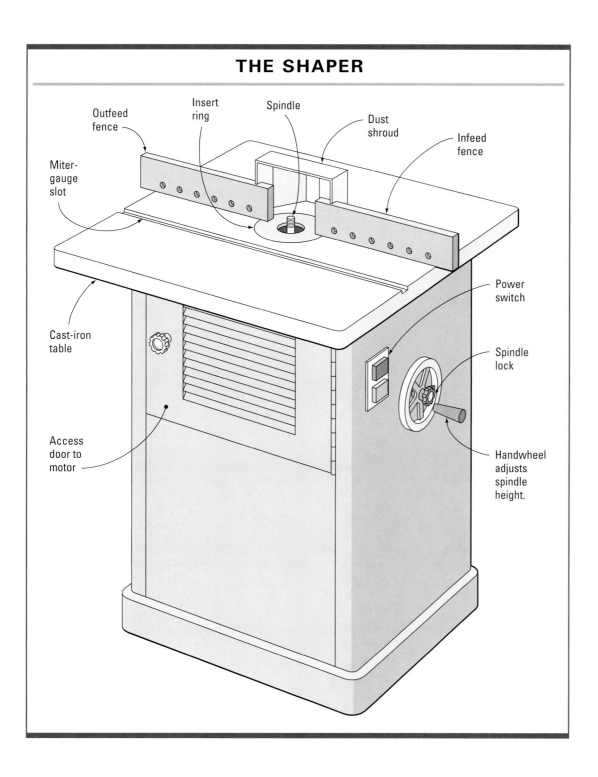

Outfeed fence

Insert ring

Spindle

Dust shroud

Infeed fence

Miter-gauge slot

Cast-iron table

Access door to motor

Power switch

Spindle lock

Handwheel adjusts spindle height.

Most shapers are equipped with a reversing switch that adds greatly to their flexibility.

meat and potatoes of the shaper, enabling it to drive large cutters through dense stock.

Typically, shapers and table-mounted routers spin counterclockwise, and the stock is fed from right to left. Yet by reversing the shaper spindle direction and inverting the cutter, you can feed the stock from left to right. This unique feature gives the shaper tremendous versatility and can make certain operations safer.

Like a table-mounted router, a shaper has a fence that guides the stock in a straight path past the cutter. The fence has independently adjustable halves and locks to the table to support the stock. For the best stock support and greatest margin of safety, the fence opening should be adjusted to be as small as possible. In fact, sometimes the best fence is one you make yourself.

This customized fence provides the greatest support to the workpiece when you're cutting rabbets.

SHAPER SAFETY

All woodworking machines are inherently dangerous, but the shaper has a reputation for being especially dangerous. Much of this stems from the complexity of the setups involved with the shaper, although the shaper is a simple machine. Woodworkers who are new to the shaper should first gain plenty of experience using a table-mounted router. Although the two machines share many of the same techniques, the table-mounted router is much smaller, less powerful, and a good machine to begin on. Also, keep in mind that shaping curves requires much more knowledge and skill than shaping straight stock. So limit yourself to shaping straight stock until you've gained plenty of experience. As one who has used a shaper for over 25 years, I can speak to the fact that this machine can be both productive and safe. Here are some additional safety guidelines I follow when I step up to the shaper:

• Read and follow the manufacturer's guidelines that come with your shaper.

• Always feed the stock against the cutter rotation. For example, when the cutter is rotating counterclockwise, feed the stock from right to left.

• Always use a guard. If the supplied guard will not work, purchase an aftermarket guard or make one of your own.

• Use safety devices, such as featherboards, push sticks, and push blocks, to help hold the work and distance your hands from the cutter.

• Take light cuts—heavy cuts create tremendous feed resistance and will often cause a kickback.

• Avoid shaping small stock. Instead, shape large stock, then cut it down to a smaller size.

• Keep the fence opening as small as possible for the greatest support of the stock.

• Disconnect the shaper from its power source when making setups.

A miter-gauge slot is milled into the top, parallel to the fence. A miter gauge provides support for shaping the ends of stock and is usually used in conjunction with the fence.

Shaper Cutters

Much of the expense in operating a shaper is in the cost of the cutters. Although a combination and rip blade are usually all that is needed for a table saw, each shaping operation requires a cutter with a unique profile. Also, a single cutter profile often costs much more than a blade for your table saw.

Fortunately, you can do a lot of shaping with a few basic profiles, a square-edged cutter, and perhaps a panel-raising cutter.

There are basically two types of shaper cutter: carbide-tipped wing cutters, and interchangeable high-speed-steel (HSS) cutters, or knives that get inserted into a cutterhead.

Wing Cutters

Like a carbide-tipped sawblade, wing cutters incorporate a steel body with separate carbide tips that are brazed permanently in place. There may be anywhere from two to

Carbide-tipped wing cutters are sharp, balanced, and smooth-running.

SHAPER CUTTERHEAD

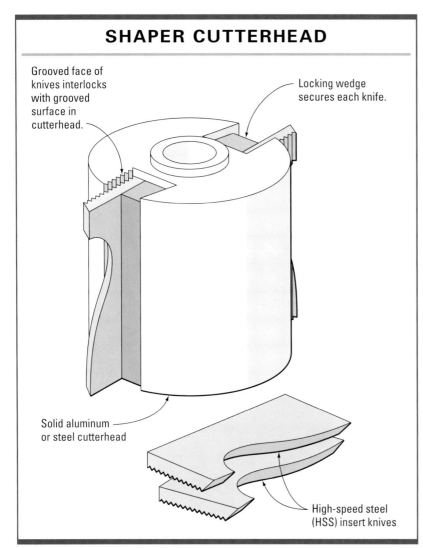

Grooved face of knives interlocks with grooved surface in cutterhead.

Locking wedge secures each knife.

Solid aluminum or steel cutterhead

High-speed steel (HSS) insert knives

five wings, but three-wing cutters are typical. Good-quality wing cutters are sharp, balanced, and extremely smooth-cutting. They're convenient, too. Just slip them over the spindle and lock them into place. Wing cutters also have the versatility for shaping straight or curved stock. Straight stock is guided by the shaper fence; for shaping curved stock, a bearing is mounted on the spindle above or below the cutter. The bearing guides the workpiece and limits the cutting depth.

Insert Cutterhead

As the name implies, an insert cutterhead uses individual knives that are inserted into the cutterhead and locked into place. The head is machined from a solid block of steel or aluminum, and the knives are secured with a mechanical interlock and machine screws. The mechanical interlock may be a pair of holes in the knives that engage with steel pins in the head, or corrugations in the knives that mate with matching corrugations in the head. Once the knives are inserted, they are locked into

POWER FEEDERS

Power feeders mount to the tops of stationary equipment to feed stock at a uniform rate. They're most useful on shapers for profiling miles of moldings but also work effectively on table saws during ripping.

Power feeders have several advantages over hand feeding. They make large-scale jobs less tiring and much more efficient; they eliminate burning and irregular mill marks associated with the inconsistencies of hand feeding; and they increase safety by keeping the operator's hands well clear of the blade or cutterhead.

Power feeders are available in a range of sizes for different applications. An induction motor drives several rubber-coated wheels through a system of gears. The gears reduce the rpm to the wheels, allow for speed changes by switching gear sizes, and provide plenty of torque without loss of power. The rubber-coated feed rollers are spring-loaded and apply pressure to the stock. The entire unit is supported by a pair of steel columns that, along with couplings that swivel and pivot, allow for universal adjustment.

The vertical column is mounted to a cast-iron base that is bolted to the machine top. Location of the base is important. You'll want to avoid obscuring the fence on the machine yet make certain that the base is close enough to the action that the feed unit will reach the stock.

Once the base is mounted, the feeder unit can be adjusted to feed either horizontally or vertically (edge or face of the stock). In fact, it can even be adjusted for feeding curved stock.

place with dovetailed wedges. Although more time-consuming to set up, insert cutterheads are an economical alternative to carbide-tipped wing cutters. While the cutterhead requires an initial investment, the profile knives are generally less expensive than a similar carbide wing cutter. And when set up correctly, insert cutterheads are safe. In fact, your jointer and planer both use a form of insert cutterheads.

One potential drawback: Unlike carbide-tipped cutters, HSS knives can't be used on man-made materials like MDF or plywood.

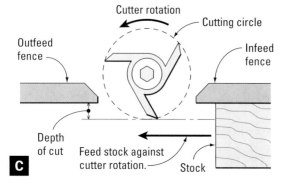

Cutter rotation

Cutting circle

Outfeed
fence

Infeed
fence

Depth
of cut

Feed stock against
cutter rotation.

Stock

Shaping Part of an Edge

The most common shaper technique is edge shaping. It's used to embellish the edges of tabletops and also for making molding strips. Narrow molding strips should be shaped on wider stock, then ripped to final size. If you're new to the shaper, edge shaping is a good place to begin (**A**).

The first step is to mount the cutter. Whenever possible, position the cutter to cut the lower edge of the workpiece (**B**). This is generally safer than cutting from the top because the cutter is buried under the workpiece. But also make note of the spindle direction—usually it's counterclockwise with the stock moving from the right side to the left (**C**). Fill the unused portion of the spindle with collars (**D**) and secure the cutter with the spindle nut (**E**). Next, adjust the height of the cutter by raising or lowering the spindle; a square positioned next to the cutter makes it easy to make a precise adjustment (**F**). Now adjust and lock the fence into place (**G**) and position a guard over the fence opening. If your shaper lacks a guard, a thick plank clamped to the fence works well (**H**). Before turning on the power, always spin the cutter by hand to make sure it clears the fence and guard. Finally, feed the stock against the rotation of the cutter (**I**).

Shaping the Entire Edge

When shaping a molding profile that cuts the entire edge **(A)**, you'll need to adjust the outfeed fence forward to compensate for the loss of stock. This is the same principle that a jointer works on: The outfeed table is set higher than the infeed table. After setting up the cutter and fence, begin by shaping 2 or 3 in. of the workpiece **(B)**, then turn the motor off. Note the gap between the molded profile and the outfeed fence **(C)**. With the power still off, use the adjustment knob on the back of the fence to move either fence so that the workpiece makes contact with the outfeed half of the fence **(D)**. Now complete the shaping process.

[TIP] **Whenever preparing stock for the shaper, always make an extra piece to be used for setup.**

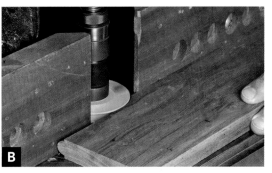

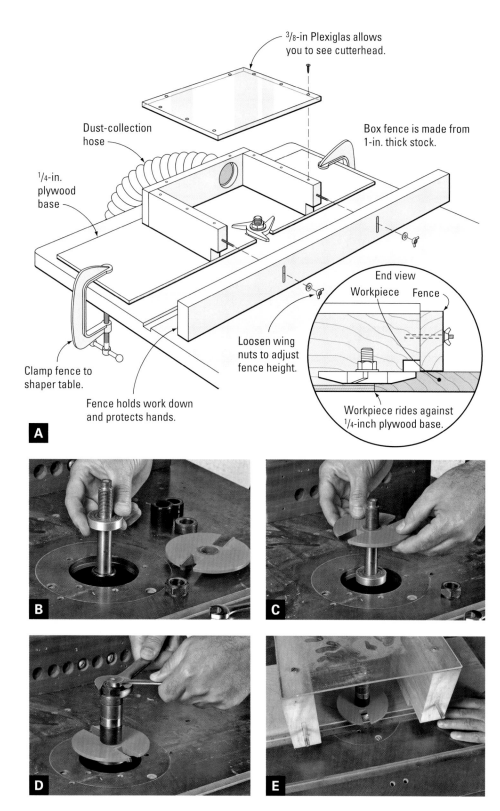

3/8-in Plexiglas allows you to see cutterhead.

Dust-collection hose

Box fence is made from 1-in. thick stock.

1/4-in. plywood base

Loosen wing nuts to adjust fence height.

End view

Workpiece   Fence

Clamp fence to shaper table.

Fence holds work down and protects hands.

Workpiece rides against 1/4-inch plywood base.

A

B

C

D

E

Raised-Panel Shaping

Frame-and-panel construction is used for doors and casework. Raised-panel cutters are large in diameter and require a large fence opening. To minimize the risks associated with a large cutter, I use a box fence that surrounds the cutter and eliminates the fence opening **(A)**.

Begin setup with a ball-bearing rub collar **(B)**. This accessory limits the depth of cut, so choose a diameter that's right for the depth of cut you want. For example, if the cutter is 3 in. in diameter and you want a 1-in. depth of cut, the rub collar should be 1 in. in diameter. Next, slip the cutter into place **(C)** and secure the spindle nut **(D)**.

Slide the box fence into position around the cutter and use a straightedge to align the fence edge with the rub collar **(E)**. When properly aligned, the edge of the fence is tangent with the rub collar **(F)**. This, in effect, creates a zero-clearance opening that dramatically increases the margin of safety because the workpiece can't be tipped or pulled into the cutter. Secure the fence to the table with a pair of clamps **(G)**. To complete the setup, attach the front of the box fence, using the workpiece to set the height **(H)**.

When you're shaping the end of a board, such as a panel, the trailing end has a tendency to splinter. The remedy is to shape the ends first **(I)**, then the edges **(J)**. The finished panel is ready for sanding and assembly within its frame **(K)**.

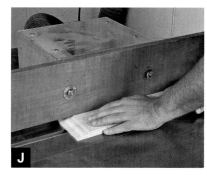

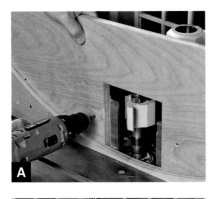

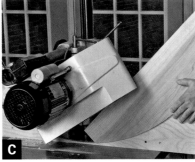

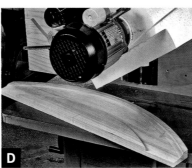

Shaping a Curved Casing

A power feeder is a great accessory for your shaper; it feeds the stock at a uniform rate, holds it firmly to the fence and table to minimize chatter, and virtually eliminates the possibility of kickback. The universal joints on the support column of a power feeder provide a wide variety of uses for this powerful tool. As you can see in this photo-essay, the power feeder can even be used for shaping curved stock, such as window casing.

Begin by setting up the shaper cutter first, then the fence. Notice that I'm using a curved "cradle" that is attached to a tall fence. Together the fence and cradle support the stock on edge as the face is shaped **(A)**.

Also, I've bandsawn the outer curve of the workpiece only; I'll saw the inside radius after shaping. This method provides increased mass and surface area, which helps reduce vibration and creates a broader surface for the feed rollers.

Next, adjust the angle of the feeder to push the workpiece downward through the arc. Also, adjust the feeder so that the spring-loaded rollers compress slightly to hold the workpiece firmly to the fence **(B)**.

Before shaping, test the setup, using the power feeder but without the shaper running, to be certain that the workpiece flows smoothly through the cradle **(C)**. Rotate the cutter so none of the wings are in the path of the workpiece. Now shape the profile. Large profiles, like this window casing, are best shaped in several passes **(D)**.

The Router Table

Basic Routing

Advanced Routing

ALTHOUGH, THE ROUTER IS A portable power tool, when a large router is mounted in a table, it becomes a mini-shaper capable of shaping both straight and curved moldings, raising panels, and even cutting joinery. While a router table lacks the size and power of a shaper, it has other important advantages. For example, the small diameter of router bits and guide bearings enables you to shape tight curves that a shaper can't. Also, many router bits—straight bits and dovetail bits, for example—cut on their ends, which enables you to cut grooves, dovetails, and other simple joints that shapers can't. Finally, don't overlook cost. You can purchase several router bits for the cost of just one shaper cutter. Clearly, the router table is a versatile tool, and an array of tables, fences, and accessories are available to add even greater versatility. Let's look at your options for getting set up with this indispensable tool.

Routers

Routers range in size from diminutive 3/4-hp laminate trimmers to heavy-duty giants with more than 3 hp and weighing around 20 lbs. Often woodworkers will begin with an average router and quickly realize its limitations. You're better off equipping your table with a large router and reserving the smaller machines for handheld use. Also, don't overlook the models with electronic variable speed, a handy feature when you're using large-diameter bits.

With a router lift, a nut recessed in the router tabletop allows you to make all height adjustments from above.

This router lift is heavy and rugged.

A router lift will allow you to make bit changes without removing the router.

This customized router table provides plenty of storage for bits, wrenches, and accessories.

Fixed-Base Versus Plunge-Base

The debate over fixed-base or plunge-base routers for table use has raged on for several years. I prefer the fixed-base for rigidity; there is typically less movement between the router motor and the base. However, clearly the best choice is a router lift. A router lift clamps to the motor and fits within the table, so the router base isn't needed. There are several advantages to a lift—it adds extra weight, reduces vibration, and makes height adjustments easy. Best of all, bit changes and precise height adjustments can be made from the tabletop without having to remove the router.

Buy the Top, Make the Table

Router tabletops are relatively inexpensive, accurately made with CNC (Computer Numerical Control) equipment, and come equipped with a miter-gauge slot. However, the best table is one you make yourself. This way, you can customize the space under the top for storage of router bits, wrenches, and other accessories.

You can easily access this power switch without having to reach under the table.

Although you can make a fence for your router table, the factory-made extruded aluminum fences have several advantages. They feature an adjustable opening, a T-slot for guards and featherboards, and dust-collection hookup.

Finally, don't overlook the power switch. Mount it where you'll have quick access.

Router Bits

Today there are more router-bit profiles available than ever before. Lots of choices mean greater flexibility and creativity, but they can also be somewhat confusing. Let's break it down.

> ## ► ROUTER TABLE SAFETY

Although most router bits are small, routers and bits deserve your respect. As with all power tools, you can enjoy them safely by following a few safety guidelines.

► See *"Shaper Safety"* on p. 239.

- Large-diameter bits are for use in a router table only. Using bits over 1 in. in diameter in a handheld router can easily cause you to lose control.

- Take light cuts. Heavy cuts invite kickback. Move the fence closer to the bit to reduce the cut, or switch to a larger pilot.

- Never climb-cut—always feed the stock from right to left.

- Distance your hands—use push sticks and push blocks to keep your hands a safe distance from the bit.

- Avoid shaping small stock—shape a larger piece and reduce it in size afterwards. If you must shape a small piece, build an appropriate jig or secure the work within the jaws of a wooden handscrew clamp.

- Always use a guard—if the fence didn't come with a guard, purchase an aftermarket guard or devise one of your own.

- Never start the router with the bit in contact with the stock.

- Don't force the bit or overload the router.

- Secure the motor in its base before starting the router.

- Don't bottom out the bit in the collet or partially insert the bit. Instead, completely insert the bit, and then back it off approximately $\frac{1}{16}$ in.

Carbide Versus High-Speed Steel

Carbide has taken over in the woodworking-tool industry. Sure, carbide is more expensive than HSS, usually around 5 to 10 times more, but it lasts 20 to 25 times longer. So it's definitely more economical in the long

► COLLET AND BIT TUNE-UP

The most important maintenance associated with your router is keeping the bits and collet clean. The collet may lose its grip on a dirty or rusty bit and spoil the cut or damage the shank of the bit. You can remove surface rust and polish the bit shanks with an abrasive pad. Dirt and pitch can build up on the collet, too, which may limit its grip. A liquid tool cleaner will dissolve the buildup on bits, collets, and other cutting tools. Finally, when inserting a router bit, don't allow the shank to bottom out against the collet. As a collet is tightened, it pulls the bit slightly downward; if the collet grips the slight radius at the base of the shank, the bit will not be held firmly.

Abrasive pads are perfect for removing gunk buildup from router-bit shanks.

Be sure to keep the router collet clean.

Use a liquid cleaner to break down stubborn buildup on bits and collets.

Don't allow bits to bottom out in the collet.

run. But you can still walk into most any hardware store and find inexpensive HSS router bits. And I still occasionally use one; they're just the thing when I need a custom profile. I can easily modify HSS bits with my bench grinder and hone the edge with slipstones. Don't try that with a carbide bit. However, for most work, carbide is the way to go. And when the bits do become dull, I drop them off at the local sharpening shop. They have the equipment and skills to handle the job.

Quarter-Inch Versus Half-Inch Shank

Hold a ¼-in. and ½-in. shank bit side by side and compare; the decision on which is better is an easy one. Half-in.-shank bits are stronger and less prone to vibration. And the greater circumference means they're less likely to slip in the collet. However, I still buy a few ¼-in. bits, mainly ¼-in. straight bits. They fit my laminate trimmer (a small, one-handed router) and make quick work of shallow mortises for hinges, locks, and other hardware.

Router-Bit Shapes

Router-bit shapes can be divided into two broad categories: joinery bits and profiling bits. Joinery bits are used to route simple joints, such as grooves and rabbets, and more complex ones, such as interlocking dovetails.

Profile bits, in contrast, cut a decorative shape, such as an ogee. Most profile bits cut basic shapes, such as beads, chamfers, and coves. Others, such as raised-panel bits, cut a profile that no other bit can duplicate.

These bits are designed for cutting joints such as grooves, rabbets, and dovetails.

Profile bits cut moldings, table edges, and other decorative shapes.

Large-diameter bits should always be run in a table-mounted router.

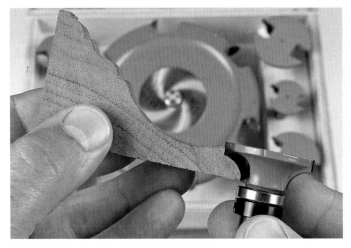

This unique bit is inverted on the shank and can shape profiles ordinary bits can't.

You can change the cutting depth of many router bits by switching to a different bearing diameter.

Piloted Versus Non-Piloted

The pilot is a tiny ball bearing that rides against the workpiece and guides the bit. On inexpensive HSS router bits, the pilot is just an extension of the bit. As the bit spins, so does the pilot. I remember the first time I used a router. It was junior high shop class, and I was using a table-mounted router equipped with an HSS cove bit to profile the edge of a small project. The pilot was naturally hot from the friction, and as it encountered a soft area of grain, it kept going. Needless to say, the scorched, irregular profile wasn't what I had in mind.

On today's carbide-tipped bits, the pilot is a tiny ball bearing that slowly rolls along the edge of the stock as the bit does its thing. It doesn't heat up, it doesn't burn, and you can even change bearing diameter to adjust the cutting depth. A pilot-guided bit is one of the best ways to shape curved work. But it's useful when shaping straight stock, too. It makes fence alignment faster and easier, and the pilot effectively reduces the size of the fence opening.

Routing an Edge

One of the most useful router techniques is to shape the edge of a tabletop or drawer front. The simple profile adds a bit of embellishment and softens the otherwise hard, square edge. You can also use this technique to create a strip of molding; first shape the edge of wider stock, then rip the molding free.

After selecting a profile and mounting the router bit, use a square to accurately adjust the height **(A)**. To set the fence, position it tangent to the guide bearing **(B)**. Before shaping, add a guard to the setup **(C)**. If shaping the entire perimeter of a board, shape the ends first, then the edges **(D)**. This technique will shape away any splintering on the corners when you're shaping the ends.

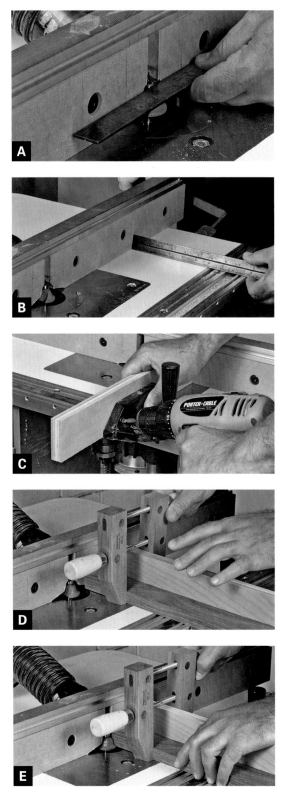

Routing End Grain

Shaping the ends of narrow stock can be a challenge; the narrow surface doesn't provide enough support along the fence. One solution is to use a miter gauge along with the fence and guide bearing on the bit.

First, set the fence tangent to the guide bearing **(A)**. Next, measure the distance from the fence to the table edge **(B)**. The distance must be equal at each end. Otherwise, the workpiece will either bind against or move away from the fence as it is fed with the miter gauge. Either way, the cut will be uneven. A backup board attached to the miter gauge will provide extra support and help avoid tearout on the trailing end of the workpiece **(C)**. To complete the setup, use a small clamp to secure the workpiece to the backup board **(D)**. Grip the clamp and the miter gauge and make the cut **(E)**.

Raising Panels

You can smoothly bevel the edges of door panels on your router table. This technique requires a large bit, so as always, make certain that the bit shank is firmly gripped by the collet. Insert the shank fully and then back it out approximately ⅛ in. **(A)**. Adjust the bit height for a light cut **(B)**; you'll want to take this cut in several passes to avoid kickback and overheating the router. Then set the fence into position tangent to the guide bearing on the router bit **(C)**. The final step of the setup is to use a guard; large bits like these are especially dangerous if left exposed **(D)**. BenchDog, Inc., makes a barrier guard especially for panel shaping **(E)**.

When shaping the panel, shape the ends first **(F)**, then the edges **(G)**. Raise the bit after each series of cuts until the full profile depth is reached.

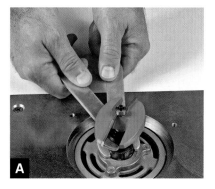

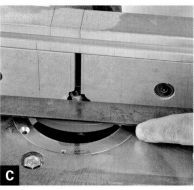

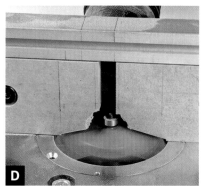

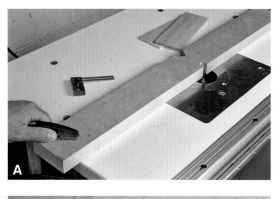

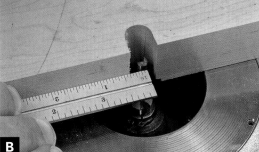

Jointing Small Stock

You can use your table-mounted router, along with an offset fence and a straight bit, to joint the edges of stock that is too small for your jointer. The fence is just a plank of hardwood with a tiny opening for a ¼-in. straight bit. After cutting the opening, use the jointer to cut ¹⁄₃₂ in. from the infeed side of the fence. This will create a ¹⁄₃₂-in. cutting depth.

Next, clamp the fence into position **(A)** so the outfeed side is tangent with the cutting edge of the bit **(B)**. As you joint the edges of the stock, keep it in contact with the outfeed side of the fence to ensure a straight edge **(C)**.

Shaping Narrow Pieces

When you need a narrow strip of molding, it's best to shape a wider board and rip the molding free. This positions your hands a safe distance from the router bit and eliminates the vibration associated with shaping narrow stock. But if

➤ See *"Shaping Part of an Edge"* on p. 242.

both edges of a narrow strip must be shaped, this method won't work. Instead, build an L-shaped jig **(A)** to hold the work, add mass, and distance your hands from the bit **(B)**.

Begin by adjusting the router-bit height **(C)** and setting the fence into position **(D)**. Then place the work in the jig and use the fence to guide the cut **(E)**. Then rotate the stock to cut the second edge **(F)**.

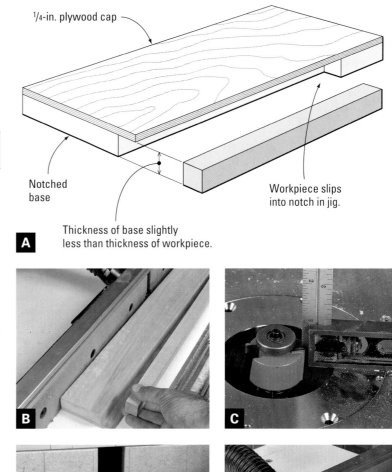

¼-in. plywood cap

Notched base

Workpiece slips into notch in jig.

Thickness of base slightly less than thickness of workpiece.

A

B

C

D

E

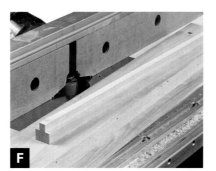

F

A

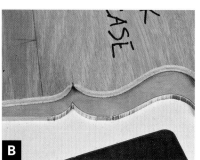

B

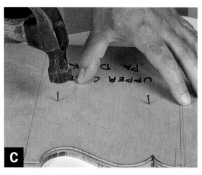

C

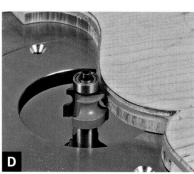

D

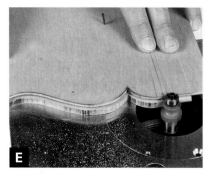

E

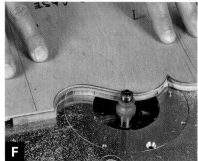

F

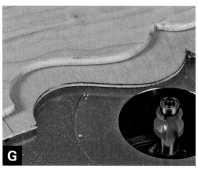

G

Template Routing

The most efficient method of shaping curves is with a template. And template shaping is the only way to shape an entire curved edge with a router. The concept is simple: A template is fastened to the workpiece to guide the bit along the contour.

Start by tracing the curves of the template onto the workpiece **(A)** and sawing the contour to remove the excess stock **(B)**. Leave between ⅟₃₂ in. and ⅟₁₆ in. of stock outside your layout lines for the router to trim. Next, attach the template to the work **(C)**. Brads work well for small profiles and light cuts; just position them in an area that won't be seen. For heavier cuts, you'll want to use screws or toggle clamps. When using screws to secure work within a template, be sure that the screws are out of the path of the router bit. Next, adjust the height of the bit to the workpiece and make certain that the bearing contacts the template **(D)**.

Now you're ready to begin shaping. Notice that the template extends beyond the workpiece **(E)**. This way, the bearing will contact the template before the bit contacts the workpiece, ensuring a safe and smooth entry into the work. As you shape, follow the contours of the template and slow the feed rate when shaping "uphill" against the grain **(F)**. Curves that meet at an inside point may need a little handwork with a chisel or sandpaper to blend the profile **(G)**.

Drilling and
Mortising Tools

Using the Drill Press

➤ Plugging Screw
Holes (p. 272)

Using the Mortiser

➤ Drilling a Mortise
(p. 273)

A LMOST EVERY WOODWORKING
PROJECT requires boring holes of
one kind or another, and many
require a number of precise holes bored to a
specific diameter and depth. For example,
when fitting a door with hardware, you'll
need to bore small holes for tiny brass
screws. Bed bolts that secure the rails to the
bedposts require large-diameter, deep holes.
Many styles of chairs, such as Windsors,
have holes in the seat for attaching the legs,
arms, and back. And if you lack a mortise
machine, you can drill a series of holes to
remove the excess stock and square the
mortise by hand with a chisel.

If you browse through the pages of a tool
catalog, you'll see a vast number of bit styles
for boring holes, as well as tools to drive the
bits. The style of bit you choose depends on
several factors: the hole diameter, the depth
of the hole, and the importance of how clean
the entry and exit of the hole needs to be. If
you're going to stop the hole at a specific
depth, the shape of the bottom of the hole
may also be important.

After boring a hole, it may be necessary to
chamfer the edges for the head of a screw, or
to plug the hole to hide the screw. There are
tools for these tasks as well. The following
pages are an overview of what's available

BENCHTOP MORTISER

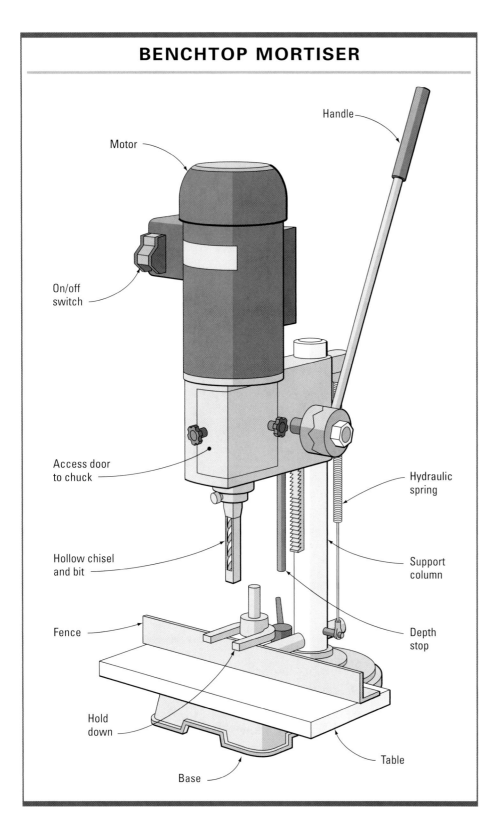

Handle

Motor

On/off switch

Access door to chuck

Hollow chisel and bit

Fence

Hold down

Base

Hydraulic spring

Support column

Depth stop

Table

Sometimes the shape of the hole bottom is important.

and of information on how to choose the best tool next time you need to bore a hole.

Twist Drills

Undoubtedly, the twist drill is the most commonly available tool for boring holes. And twist drills are easy to get your hands on, too. In fact, you can find them in almost any woodworking catalog, home center, or local hardware store. They're readily available in sizes from $\frac{1}{16}$ in. to $\frac{1}{2}$ in. Best of all, twist drills easily cut both wood and metal. So besides using them to bore holes in wood, you can use them around the shop to modify brass hardware or attach an accessory power feed to the cast-iron top of a machine. Twist drills cut cleanly and quickly, too. Although they don't cut the cleanest entry and exit hole, they will do an acceptable job, especially when freshly sharpened. However, as twist drills become dull, the quality of the edges of the hole drops significantly. Also, twist drills have an annoying tendency to wander when you're starting a hole, so it's a good idea to mark the location of the hole with an awl or

A punch dimples the surface, which helps to prevent twist-drill bits from wandering.

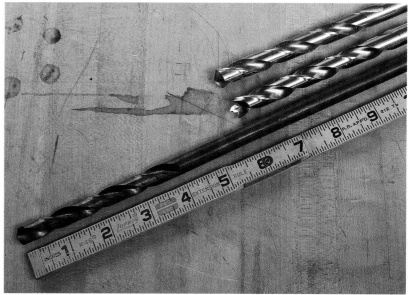

The extra length of electricians' bits will enable you to bore deep holes—for example, when drilling for bed bolts.

Brad-point bits are designed specifically for woodworking.

punch before drilling. For drilling deep holes, twist drills are available in long lengths; you can usually find these tools at hardware stores or wherever electricians' tools are sold.

Brad-Point Bits

Brad-point bits are designed exclusively for boring holes in wood. At first glance, a brad-point bit looks like an ordinary twist drill. The difference is in the point and spurs, which serve to accurately locate the hole and cleanly shear the wood that forms the hole's edges. Good-quality brad-point bits tend to be somewhat expensive, so I reserve their use for when the edges of the hole must be

clean. Both twist-drill bits and brad-point bits will stop cutting when the flutes become clogged, and small-diameter bits clog more easily than large ones. When boring a deep hole, withdraw the bit occasionally to allow the chips to escape. It's also helpful to coat the surface of the bit with a woodworking tool lubricant before you begin drilling. Brad-point bits are commonly available in sizes from $1/4$ in. to $1/2$ in.

Spade Bits

As the name implies, spade bits resemble spades. These simple tools are widely available, inexpensive, and bore a remarkably clean hole. An improved spade bit that has recently hit the market features spurs to cleanly shear the fibers for even better results. Spade bits are also available in extra-long lengths for boring deep holes. The point of a spade bit isn't always ground to be

Spade bits are inexpensive and will bore a fairly clean hole.

perfectly concentric with the shank, so it's good practice to mark the center of the hole with an awl before beginning. Also, the point of a spade bit is long, so this tool may not be the best choice for stopped holes. Even so, spade bits cut cleanly and quickly, and the price is within easy reach. As an added advantage, spade bits are easy to modify. For example, if you need a tapered or slightly undersized hole, you can carefully grind the edges of the bit. Spade bits range in size from $\frac{1}{4}$ in. to $1\frac{1}{2}$ in.

Forstner Bits

Forstner bits are unique among wood-boring bits. While most bits are guided by their centers, a Forstner bit is guided by its rim. Because of its unusual design, the Forstner bit can bore overlapping holes. If you try this with other styles of bits, they'll wander into the adjacent hole. The sharp rim also cleanly

The improved version of the spade bit has a scoring spur so it will cut a cleaner entry than ordinary spade bits do.

Forstner bits cut amazingly sharp holes and create a flat bottom.

Forstner bits are guided by their edge and can cut clean, overlapping holes.

Augers have a square taper on the shank to fit into a bit brace.

Augers are useful for boring large holes or angled holes.

shears the wood fibers to create a very precise entry. It's also the only bit that bores an absolutely flat-bottomed hole. Because Forstner bits tend to be expensive and difficult to sharpen, I reserve their use for times when the work requires a clean, precise hole. Let's take a look at some other useful bits.

Augers

You may have a collection of augers left to you by your grandfather; these old tools have been around a while. Despite their dated design, they're still serviceable at times. I use augers for boring large holes (those above 1/2 in. in diameter), especially holes bored at angles other than 90 degrees. Augers are hand driven with a brace so you can bore the hole slowly and carefully. The point of an auger has a lead screw that bites into the wood and actually pulls the auger into the

The lead screw of an auger pulls the bit through the wood.

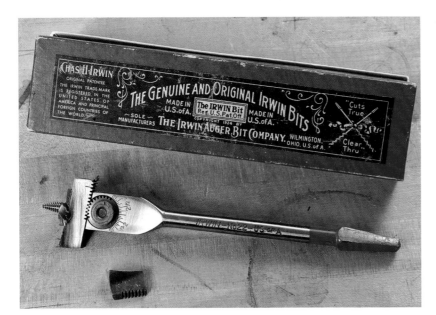

An expansion bit can be useful for drilling odd-sized holes.

The spur of an auger slices the wood with a wedging action.

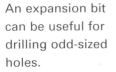

tools, they don't work quite as well as ordinary augers. With only one spur the expansion bit cuts slowly, and boring angled holes is difficult at best. And although it will cut holes of various sizes, it's tedious to set up for a precise diameter. Even so, an expansion bit can be useful for an occasional odd-sized hole.

Spoon Bits

Spade bits resemble spades; spoon bits look like—you guessed it—spoons. This is another old-style bit that is still quite useful today. The unique shape of the spoon bit leaves a round bottom hole that is ideal for chairmaking. When you bore holes in a chair leg to receive a tenon from the stretcher, the hole can weaken the leg. This is especially true if the bottom of the hole has sharp, square edges, like those from a Forstner bit. Stresses on the completed chair can focus on the acute, inside corner of the hole. The spoon bit alleviates the problem by creating a round-bottomed

stock as you bore the hole. The rim of an auger has two spurs that cleanly sever the wood at the rim of the hole.

Expansion bits are probably best described as adjustable augers. They feature one spur rather than two, and it slides into the head of the bit and locks into place with a single screw. Like most 19th-century multipurpose hole. Like augers, spoon bits have tapered, square tangs that are designed to fit within the jaws of a brace.

Reamers

To shape a long taper into an existing hole, the reamer is the tool of choice. This bit actually shaves the side of a hole to a taper. The taper can accommodate a round, tapered tenon on a chair leg. This technique is used to construct a Windsor-style chair, in an ingenious method of construction that causes the leg of the chair to be driven tighter into the seat each time someone sits down. Tapers have long been known to provide incredibly secure joints. For example, a drive center in a lathe fits to the headstock with a taper; it requires a knockout rod to remove it. Reamers are available for both the hand brace as well as an electric drill.

A reamer is used to taper a hole to fit a tapered leg.

Countersinks

The word "countersink" is both a noun and a verb. It refers to a shallow chamfered opening around a hole in which to fit the head of a screw; it also refers to the tool used to create it. Countersinks are available with a single flute or multiple flutes. Surprisingly, a single-flute countersink cuts more smoothly than the multi-flute style, which has a tendency to chatter. A unique

The countersink bevels the edge of a hole to fit the head of a screw.

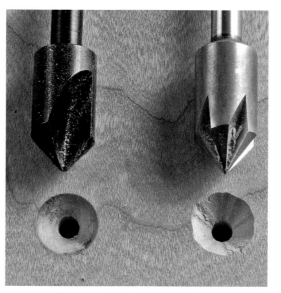

Single-flute countersinks cut more smoothly than multi-flute types.

This bit drills and countersinks in one step.

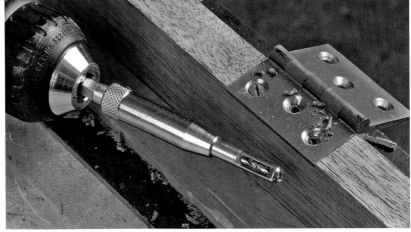

This bit centers itself when you're drilling for hardware.

The tip of the tool aligns the bit with the existing hole in a hinge or other hardware.

A center punch makes drilling holes more accurate.

and very handy style of countersink both drills a hole and countersinks it with one bit.

Another specialty drilling tool is the self-centering bit. This neat device makes drilling for hardware a snap. Simply position the tool in the mounting holes of a hinge and drill. The hole will be perfectly centered—every time.

To accurately position large holes on center, I use a center punch. The hardened, conical tip produces a dimple in wood or metal to keep bits from wandering as the hole is started.

Drill Stops and Guides

Often it's necessary to stop drilling at a specific depth. Whether you're boring large holes for a bed bolt or small holes for tiny brass screws, it's usually not necessary or even desirable to bore completely through. Although you can purchase stop collars that slip over the bit and clamp into place, I find that a small wood block works as well. If the exact depth of the hole is not critical, I

A block of wood works well for a depth stop.

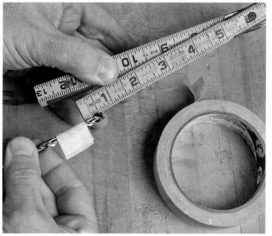

Another simple depth stop is a piece of masking tape.

just stick a piece of tape to the bit to show where to stop.

For drilling tiny holes for escutcheon pins or small nails, a brad works well as a drill. Just snip off the head and use it as you would a twist drill. Technically speaking, the brad isn't removing wood as it creates a hole; it is only compressing the fibers. But this simple method works surprisingly well for small-diameter holes, and the nail is less likely to break than a small bit is.

If you would like to drill a perfectly perpendicular hole by hand, then consider a drill guide. Some have a framework that supports a power drill, while others use hardened steel bushings to support the bit.

For a tiny drill bit, just snip the head off a brad.

Plug Cutters

Using a screw to join wood certainly isn't sophisticated joinery, but it works well in some situations and may even be the best option. To hide the head of the screw you

A drill guide will enable you to bore accurate holes without a drill press.

can cover it with a wooden plug. The best plug cutters create a small taper on the walls of the plug so that it fits snugly within the hole. A plug cutter works best in a drill press; I use the stop on the drill press so the plugs are equal in length.

A tapered plug cutter produces face-grain plugs that are nearly invisible after finishing.

For power and convenience, it's tough to beat a cordless drill.

Driving the Bits

Although drilling a hole is certainly not a demanding task, it's still helpful to have the right tool for the job. Power drills have become sophisticated with features such as keyless chucks, variable speed, reverse, and even adjustable clutches for controlled power driving of screws. Best of all, they're cordless. The cordless drill is my favorite tool for boring a hole. It's fast, powerful, and convenient.

For speed and accuracy, it's tough to beat a drill press. With a drill press, you'll get exact, perpendicular holes. The drill press also has a built-in depth stop. Although the shank of some bits is turned down to fit a $^3/_8$-in. chuck, it's safer to bore large holes in a drill press. Bits with tapered, square tangs are designed specifically for a bit brace.

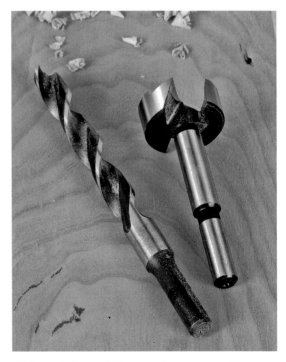

Although large bits may have a small shank, it's best to use them in a drill press.

▶ DRILL PRESS OPTIONS

A drill press has several advantages over a portable drill for boring holes. The induction motor has much greater torque so you can bore larger holes; the holes will be perpendicular to the face of the stock; and the built-in stop of a drill press insures that the depth of each hole is accurate and consistent.

The drill press is stone simple, a motor driving a drill chuck via a belt and pulleys. All drill presses have multiple-speed step pulleys. A table supports the stock; as the lever is pulled, the chuck advances the spinning bit.

There are three types of drill presses from which to choose—floor models, radial models, and benchtop models. Of the three, I prefer a benchtop drill press. I built a storage cabinet for mine; the rows of drawers utilize space wasted by a floor-model press. The radial-model drill press allows you to drill angled holes, but the setup is usually impractical; I find it much easier to drill an angled hole with a portable drill.

A drill press is the most accurate tool for boring holes.

Bits with a square, tapered shank are designed for use in a bit brace.

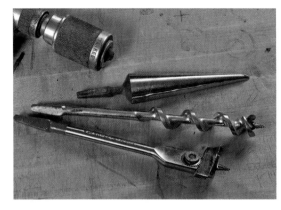

The old eggbeater-type hand drill can still be useful for an occasional small-diameter hole.

This is a compact version of the freestanding mortiser. Benchtop models are a good choice for most shops.

Although the bit brace has been around for centuries, the much-improved modern brace has a ratcheting chuck and can be reversed if necessary.

Eggbeater drills can still be useful for drilling small-diameter holes, and kids enjoy using this simple, mechanical tool. Just make sure to give them a scrap block for experimenting.

The Mortiser

The mortiser is a beefy little drill press that bores square holes. Boring a row of overlapping holes creates a mortise. Years ago, if you wanted a mortise machine, you could choose between an expensive industrial model, which consumed more than its fair share of floor space, or an attachment for your drill press. The drill-press attachment was time-consuming to install, and, of course, you couldn't bore round holes without removing it. Hardly an efficient method. I continued to cut mortises by hand until the benchtop mortisers were introduced. Although many benchtop machines lack the size and muscle of their stationary cousins, this isn't true with

A mortise is created by boring a row of overlapping holes.

benchtop mortisers. The better examples are rugged and powerful and have plenty of capacity for most mortises in furniture. In fact, some manufacturers have even developed compact floor models with even greater power and capacity.

Mortiser Anatomy

Mortise machines bore holes with a round bit that fits within a square, hollow chisel. As this assembly is forced into the workpiece, the bit bores a round hole as the chisel squares the corners. One side of the chisel is open to allow the chips to escape.

The chisel fits firmly in a machined chuck and is held secure with a setscrew. The bit fits within the jaws of a drill chuck that is mounted directly to the shaft of the motor. As the feed lever is pulled, the entire assembly—bits, motor, and chucks—is lowered along a vertical rack, or gear. A cast-iron column supports the rack, motor, and boring assembly.

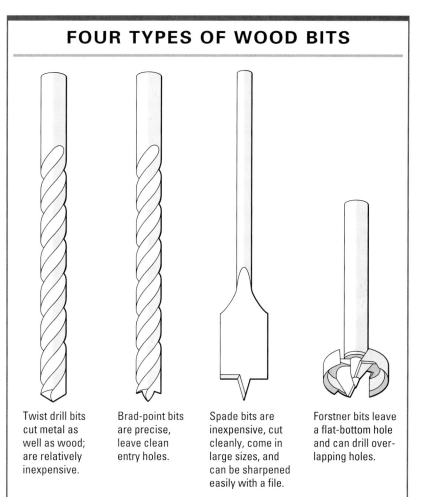

FOUR TYPES OF WOOD BITS

Twist drill bits cut metal as well as wood; are relatively inexpensive.

Brad-point bits are precise, leave clean entry holes.

Spade bits are inexpensive, cut cleanly, come in large sizes, and can be sharpened easily with a file.

Forstner bits leave a flat-bottom hole and can drill overlapping holes.

A table supports the workpiece during mortising. Some tables have an integral fence. To set up the location of the cut, the table and fence move front to back. On many benchtop mortisers, the table and fence are separate units, and setups are made by moving the fence fore and aft. A most important feature is the depth stop, which limits the travel of the head so that all cuts are of the same depth.

A

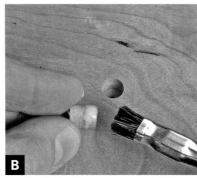

B

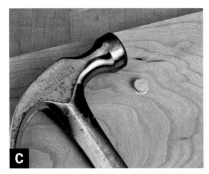

C

D

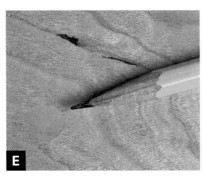

E

Plugging Screw Holes

To cut matching plugs for screw holes, chuck a plug cutter into the drill press, set the drill-press stop, and drill a series of plugs. Make sure that the plugs are longer than the holes are deep **(A)**. Snap the plugs free with a screwdriver. After selecting a plug with grain similar to that of the wood surrounding the hole, dab a light coat of glue on the plug **(B)** and tap it into place **(C)** with a hammer. Remember to align the grain for a good match. After trimming the excess with a flush-trimming saw **(D)**, smooth the surface with a block plane **(E)**.

Drilling a Mortise

You can cut a mortise by hand with a chisel and mallet. But a mortise machine speeds the process along. Begin by marking the mortise location **(A)**, then use a marking gauge to scribe the mortise walls **(B)**.

➤ See *"Mortise-and-Tenon Joint"* on p. 115.

After setting the mortise location and depth on the machine, begin by cutting the mortise at each end **(C)**. This method ensures that the ends of the mortise are square. The mortise chisel will bore true when cutting on two sides or four. But if the chisel cuts on three sides, as it does with overlapping holes, it will have a tendency to slip into the adjacent hole. Next, starting at one end, stagger the cuts **(D)**. Finally, make a second series of cuts to mortise out the remaining material **(E)**.

Further Reading

SETTING UP

Bird, Lonnie. *The Bandsaw Book.*
Taunton Press.
Sandor, Nagyszalanczy. *Woodshop Dust Control.*
Taunton Press.

SHARPENING

Darlow, Mike. *Woodturning Chucks and Jigs.*
Melaleuca Press.
Lee, Leonard. *The Complete Guide to Sharpening.*
Taunton Press.

WOOD

Alexander, John. *Making a Chair from a Tree.*
Taunton Press.
Hoadley, Bruce. *Understanding Wood.*
Taunton Press.
Malloff, Will. *Chainsaw Lumbermaking.*
Taunton Press.
O'Donnell, Michael. *Turning Green Wood.*

WOODTURNING TECHNIQUE

Mortimer, Stuart. *Techniques of Spiral Work.*
Lyons and Burford.
Raffan, Richard. *Turning Wood with Richard Raffan.*
Taunton Press.
Raffan, Richard. *Turning Bowls with Richard Raffan.*
Taunton Press.
Raffan, Richard. *Turning Boxes with Richard Raffan.*
Taunton Press.

THREAD CHASING

Darlow, Mike. *Woodturning Techniques.*
Melaleuca Press.
Jacob, John. *Hand or Simple Turning.*
Holtzapffel, Dover Publications.

FINISHES

Finishes and Finishing Techniques. Taunton Press.
Dresdner, Michael. *The New Wood Finishing Book.*
Taunton Press.

Index

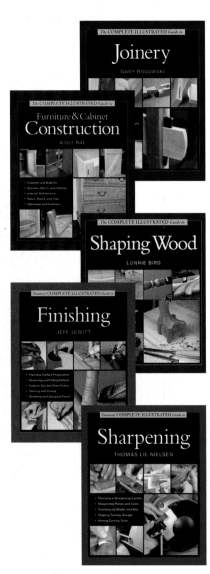

Other Books in the Series:

HARDCOVER

The Complete Illustrated Guide to Joinery
Gary Rogowski
ISBN 1-56158-401-0
Product #070535
$39.95

The Complete Illustrated Guide to Furniture and Cabinet Construction
Andy Rae
ISBN 1-56158-402-9
Product #070534
$39.95

The Complete Illustrated Guide to Shaping Wood
Lonnie Bird
ISBN 1-56158-400-2
Product #070533
$39.95

Taunton's Complete Illustrated Guide to Finishing
Jeff Jewitt
ISBN 1-56158-592-0
Product #070712
$39.95

Taunton's Complete Illustrated Guide to Sharpening
Tom Lie-Nielson
ISBN 1-56158-657-9
Product #070737
$39.95

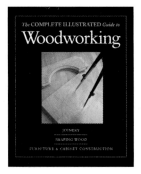

THE COMPLETE ILLUSTRATED GUIDES SLIPCASE SET

The Complete Illustrated Guide to Joinery

The Complete Illustrated Guide to Furniture and Cabinet Construction

The Complete Illustrated Guide to Shaping Wood
ISBN 1-56158-602-1
Product #070665
$120.00

THE COMPLETE ILLUSTRATED GUIDES SLIPCASE SET

Taunton's Complete Illustrated Guide to Using Woodworking Tools

Taunton's Complete Illustrated Guide to Sharpening

Taunton's Complete Illustrated Guide to Finishing
ISBN 1-56158-745-1
Product #070817
$126.00

PAPERBACK

Taunton's Complete Illustrated Guide to Period Furniture Details
Lonnie Bird
ISBN 1-56158-590-4
Product #070708
$27.00

Taunton's Complete Illustrated Guide to Choosing and Installing Hardware
Bob Setttich
ISBN 1-56158-561-0
Product #070647
$29.95

Taunton's Complete Illustrated Guide to Box Making
Doug Stowe
ISBN 1-56158-593-9
Product #070721
$24.95